Tania Tronco (Ed.)

New Network Architectures

Studies in Computational Intelligence, Volume 297

Editor-in-Chief
Prof. Janusz Kacprzyk
Systems Research Institute
Polish Academy of Sciences
ul. Newelska 6
01-447 Warsaw
Poland
E-mail: kacprzyk@ibspan.waw.pl

Further volumes of this series can be found on our homepage: springer.com

Vol. 276. Jacek Mańdziuk
Knowledge-Free and Learning-Based Methods in Intelligent Game Playing, 2010
ISBN 978-3-642-11677-3

Vol. 277. Filippo Spagnolo and Benedetto Di Paola (Eds.)
European and Chinese Cognitive Styles and their Impact on Teaching Mathematics, 2010
ISBN 978-3-642-11679-7

Vol. 278. Radomir S. Stankovic and Jaakko Astola
From Boolean Logic to Switching Circuits and Automata, 2010
ISBN 978-3-642-11681-0

Vol. 279. Manolis Wallace, Ioannis E. Anagnostopoulos, Phivos Mylonas, and Maria Bielikova (Eds.)
Semantics in Adaptive and Personalized Services, 2010
ISBN 978-3-642-11683-4

Vol. 280. Chang Wen Chen, Zhu Li, and Shiguo Lian (Eds.)
Intelligent Multimedia Communication: Techniques and Applications, 2010
ISBN 978-3-642-11685-8

Vol. 281. Robert Babuska and Frans C.A. Groen (Eds.)
Interactive Collaborative Information Systems, 2010
ISBN 978-3-642-11687-2

Vol. 282. Husrev Taha Sencar, Sergio Velastin, Nikolaos Nikolaidis, and Shiguo Lian (Eds.)
Intelligent Multimedia Analysis for Security Applications, 2010
ISBN 978-3-642-11754-1

Vol. 283. Ngoc Thanh Nguyen, Radoslaw Katarzyniak, and Shyi-Ming Chen (Eds.)
Advances in Intelligent Information and Database Systems, 2010
ISBN 978-3-642-12089-3

Vol. 284. Juan R. González, David Alejandro Pelta, Carlos Cruz, Germán Terrazas, and Natalio Krasnogor (Eds.)
Nature Inspired Cooperative Strategies for Optimization (NICSO 2010), 2010
ISBN 978-3-642-12537-9

Vol. 285. Roberto Cipolla, Sebastiano Battiato, and Giovanni Maria Farinella (Eds.)
Computer Vision, 2010
ISBN 978-3-642-12847-9

Vol. 286. Zeev Volkovich, Alexander Bolshoy, Valery Kirzhner, and Zeev Barzily
Genome Clustering, 2010
ISBN 978-3-642-12951-3

Vol. 287. Dan Schonfeld, Caifeng Shan, Dacheng Tao, and Liang Wang (Eds.)
Video Search and Mining, 2010
ISBN 978-3-642-12899-8

Vol. 288. I-Hsien Ting, Hui-Ju Wu, Tien-Hwa Ho (Eds.)
Mining and Analyzing Social Networks, 2010
ISBN 978-3-642-13421-0

Vol. 289. Anne Håkansson, Ronald Hartung, and Ngoc Thanh Nguyen (Eds.)
Agent and Multi-agent Technology for Internet and Enterprise Systems, 2010
ISBN 978-3-642-13525-5

Vol. 290. Weiliang Xu and John E. Bronlund
Mastication Robots, 2010
ISBN 978-3-540-93902-3

Vol. 291. Shimon Whiteson
Adaptive Representations for Reinforcement Learning, 2010
ISBN 978-3-642-13931-4

Vol. 292. Fabrice Guillet, Gilbert Ritschard, Djamel A. Zighed, and Henri Briand (Eds.)
Advances in Knowledge Discovery and Management, 2010
ISBN 978-3-642-00579-4

Vol. 293. Anthony Brabazon, Michael O'Neill, and Dietmar G. Maringer (Eds.)
Natural Computing in Computational Finance, 2010
ISBN 978-3-642-13949-9

Vol. 294. Manuel F.M. Barros, Jorge M.C. Guilherme, and Nuno C.G. Horta
Analog Circuits and Systems Optimization based on Evolutionary Computation Techniques, 2010
ISBN 978-3-642-12345-0

Vol. 295. Roger Lee (Ed.)
Software Engineering, Artificial Intelligence, Networking and Parallel/Distributed Computing, 2010
ISBN 978-3-642-13264-3

Vol. 296. Roger Lee (Ed.)
Software Engineering Research, Management and Applications, 2010
ISBN 978-3-642-13272-8

Vol. 297. Tania Tronco (Ed.)
New Network Architectures, 2010
ISBN 978-3-642-13246-9

Tania Tronco (Ed.)

New Network Architectures

The Path to the Future Internet

Tania Tronco
CPqD Telecom & IT Solutions
Directory of Technology Innovation
Rod. Campinas/Mogi-Mirim, km118,5
13086-905 - Campinas - São Paulo
Brazil
E-mail: tania@cpqd.com.br

ISBN 978-3-642-13246-9   e-ISBN 978-3-642-13247-6

DOI 10.1007/978-3-642-13247-6

Studies in Computational Intelligence   ISSN 1860-949X

Library of Congress Control Number: 2010929356

© 2010 Springer-Verlag Berlin Heidelberg

This work is subject to copyright. All rights are reserved, whether the whole or part of the material is concerned, specifically the rights of translation, reprinting, reuse of illustrations, recitation, broadcasting, reproduction on microfilm or in any other way, and storage in data banks. Duplication of this publication or parts thereof is permitted only under the provisions of the German Copyright Law of September 9, 1965, in its current version, and permission for use must always be obtained from Springer. Violations are liable to prosecution under the German Copyright Law.

The use of general descriptive names, registered names, trademarks, etc. in this publication does not imply, even in the absence of a specific statement, that such names are exempt from the relevant protective laws and regulations and therefore free for general use.

Typeset & Cover Design: Scientific Publishing Services Pvt. Ltd., Chennai, India.

Printed on acid-free paper

9 8 7 6 5 4 3 2 1

springer.com

Preface

The Internet has invaded most aspects of life and society, changing our lifestyle, work, communication and social interaction and giving us expectations about new forms of interactions, access to global knowledge and decrease of the digital divide. Nevertheless, the current Internet suffers with lack of mobility, loss of transparency, scalability problems, incompatibility among protocols, protocols taking roles for which they weren't originally designed, security vulnerability and attacks. Nowadays it is taking place a big momentum on Future Internet research. It is time to rethink the Internet architecture and reengineering it to address the current and future requirements. There is a common consensus that the Internet needs improvement. Nevertheless, there is not yet a shared vision on how this may happen. There is not a complete network science to accurately predict and control network behaviors with global interactions.

It is the aim of this book to group and to describe in a concise way diverse new Internet architecture proposals and ongoing research projects to readers have an overview and better understanding what is happening now in future Internet research area. A survey about the evolution of the Internet architecture, its principles and a brief history of the Internet are also presented to illustrate why Internet architecture needs to change.

The idea for this book came with an invitation from Springer during the time of the CPqD Workshop on New Architectures for Future Internet, occurred in 23-24 September, 2009, in the context of the ARCMIP (ARchitectures for Mobile IP), future Internet project sponsored by FUNTTEL (Funding for Technological Development of the Telecommunications) - Ministry of Communications, Brazil. The workshop took place at CPqD Foundation, the largest ICT R&D Center in Brazil, at Campinas, São Paulo, where ARCMIP project is being developed. It consisted of a number of selected presentations from key national and international researchers that sharing their experiences with us and some of them accepted in contribute to create a book about the workshop's theme including their presentations.

I would like to thank all the authors for their ready availability to write despite their intense research activities, CPqD Foundation, Antonio Marcos Alberti, Christian Esteve Rothenberg, Takashi Tome, Mayra Castro, Luis F. de Avila for the illustrations and all the people that some way contributed to the creation of this book.

Campinas, SP   Tania Regina Tronco
May, 2010

Contents

A Brief History of the Internet 1
Tania Regina Tronco

Principles of Internet Architecture 13
Tania Regina Tronco

Evolution of Internet Architecture 25
Tania Regina Tronco

Scenarios of Evolution for a Future Internet Architecture 57
Tania Regina Tronco, Takashi Tome, Christian E. Rothenberg, Marco A. Ongarelli, A.C. Bordeaux Rego

Future Network Architectures: Technological Challenges and Trends 79
Antônio Marcos Alberti

New Generation Internet Architectures: Recent and Ongoing Projects 121
Tania Regina Tronco, Takashi Tome, Christian E. Rothenberg, Antonio Marcos Alberti

OneLab: An Open Federated Facility for Experimentally Driven Future Internet Research 141
Serge Fdida, Timur Friedman, Thierry Parmentelat

RNP Experiences and Expectations in Future Internet Research and Development 153
Michael Stanton

Description of Network Research Enablers on the Example of OpenFlow 167
Julius Werner

Re-architected Cloud Data Center Networks and Their Impact on the Future Internet 179
Christian Esteve Rothenberg

Improving the Scalability of Internet Routing 189
Christian Vogt

Delivering Building Blocks for Internet of Services: Trust, Security, Privacy and Dependability 205
Aljosa Pasic

ITU Focus Group on Future Networks 215
Nilo Pasquali, Abraão B. Silva

Key Issues on Future Internet 221
Tereza Cristina Melo de Brito Carvalho, Charles Christian Miers, Cristina Klippel Dominicini, Fernando Frota Redígolo

Challenges for the Brazilian Green and Yellow Internet 237
Djamel F.H. Sadok, Judith Kelner, Joseane Fidalgo

Author Index 249

A Brief History of the Internet

Tania Regina Tronco

CPqD Foundation, Rodovia Campinas Mogi-Mirim, km 118,5,
Campinas – São Paulo, CEP 13096-902, Brazil
tania@cpqd.com.br

Abstract. This chapter introduces a brief history review of Internet with focus on its original conception. It's important to remember such initial ideas because they were the basis of Internet architecture, they are still at the core of today's Internet and they can be helpful to rethink new design requirements nowadays. Hence, we start by the initial packet-based network protocols and their evolution to TCP/IP.

1 Introduction

The Internet architecture concept was conceived at the end of the 60´s by ARPA (Advanced Research Projects Agency) during the Cold War, when the United States and Soviet Union were preparing for an eventual military confrontation. At that time, the U.S. military created an underground network of cables and equipments intended to survive a nuclear attack. This network was named ARPANET and its design consisted of a number of requirements such as:

- Data should be moved through leased lines to avoid problems with interruptions of the telephone system;
- The information to be transmitted should be broken into segments of fixed length (packets) instead of being a continuous stream and
- The network should be totally decentralized, without a single node in the control of the network, yielding reliability and robustness.

ARPANET was opened to universities after the end of arms race and a key requirement was added to the network project:

- Communication between computers, called hosts, should be done through devices called Interface Message Processor (IMP), as shown in Fig. 1.

The IMP function was to receive messages from a host and break them in packets. These packets should pass from IMP to IMP through the network until the destination IMP, which should pass them to the destination host.

The network consisted of the interconnection of these IMPs through the leased lines supplied by telephonic companies. The first IMP was built by the company

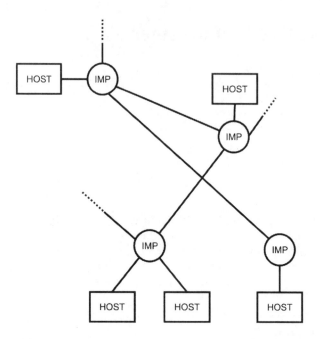

Fig. 1 A Typical Section of ARPANET (adapted from [1])

BBN (Bolt Beraneck and Newman) from Cambridge in 1976. The report No. 1822 of BBN [2] contains the specifications for the interconnection of a host and an IMP. According to this report, for each regular message, the host specified a destination, composed of three parameters: IMP, host and handling type. These parameters specified uniquely a connection between source and destination host. The handling type was used to specify characteristics of the connection, such as priority or non-priority of transmission. The messages should be sent to the destination in the same order that were transmitted by the source and, for each regular message, the host also specified a 12 bit identifier to be used with the destination of the message, forming a *message-id*, in order to retransmit them in case of the network failure.

The first IMP was installed at University of California (UCLA), in Los Angeles, followed by SRI (Stanford Research Institute), University of California in Santa Barbara and University of Utah, 4 points in total. The first ARPANET transmission was made between UCLA and SRI in Mento Park, California in 1969. In the same year, the first RFC (Request for Comments) was published; RFC3 defined the RFC series for ARPANET and later, the Internet.

2 Decade of the 70´s

After installing some IMPs in a network, the objective of DARPA was to standardize the ARPANET network interface to allow more DARPA sites to join the network. To achieve this, the first standard networking protocol was developed in December 1970, namely Network Control Protocol (NCP) [4].

2.1 Network Control Protocol Operation

The NCP operation consisted of store-and-forward messages from a sending host to a receiving host. After a host sent a message, it was prohibited from sending another message until receives a RFMN (Request-for-Next-Message). This sequence of requests made a connection. A connection linked two processes between a sending and a receiving host.

The primary function of the NCP was to establish connections and release connections. In order to send control commands to establish and release connections between the hosts, one particular link, designated as the control link, was established between each pair of host.

Each host had its internal naming scheme, often incompatible with other hosts. Then, an intermediate name space, named socket, was created in NCP to prevent using this internal name scheme. Each host was responsible for mapping its inner process identifiers into sockets as shown in shown in Fig. 2.

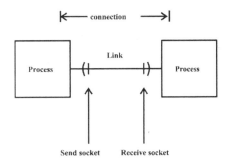

Fig. 2 A Typical Socket (adapted from [4])

A socket specifies one connection endpoint and is determined by three numbers:

- **A user number** (24 bits) composed by:
    - 8-bit for home host number,
    - 16 bits to identify him at that host.
- **A host number** (8 bits)
- **An AEN** (Another Eight-bit Number) composed by:
    - 1 bit that indicate a receive host (=0) or a send host (=1);
    - 7 bits – that provide a population of 128 sockets for each used number at each host.

When a user tried to log into a host, her user number was used to tag all the processes created in that host, producing a sort of virtual network.

By the end of 1971, there were fifteen sites attached into ARPANET using NCP [10] as follows:

- Bolt Baranek and Newman (BBN)
- Carnegie Mellon University
- Case Western Reserve University
- Harvard University
- Lincoln Laboratories
- Massachusetts Institute of Technology (MIT)
- NASA at AMES
- RAND Corporation
- Stanford Research Institute (SRI)
- Stanford University
- System Development Corporation
- University of California at Los Angeles (UCLA)
- University of California of Santa Barbara
- University of Illinois at Urbana
- University of UTAH

At this time, BBN also developed an electronic mail program for ARPANET that quickly became the most popular application on the ARPANET [11]. The e-mail program specified the destination address as username@hostname, where username was the same used to login in the host.

At the end of the seventies, there were about 200 hosts connected to ARPANET [11]. The NCP was becoming inefficient to connect different packet switching networks because individual networks could differ in their implementations like the heterogeneous addressing schemes, the different maximum size for the data, the different time delays for accepting, delivering, and transporting data and so on. In May 1974, Robert Kahn and Vinton Cerf published a paper entitled "A Protocol for Packet Network Intercommunication" on IEEE Transaction on Communication [3], proposing a new protocol to support the sharing resources between different packet switching networks. This protocol was named TCP (Transmission Control Protocol).

According to [3], for both economic and technical considerations, it was convenient that all the differences between networks could be resolved by simple and reliable interface. This interconnection should also preserve intact the internal operation of each individual network. This interface was named Gateway.

Fig. 3 illustrates two network interconnected by one gateways.

The gateway was divided into two parts; each one associated with its own network and its function was understand the source and destination host addresses and insert this information in a standard format in every packet. For this operation, an internetwork header was added to the local header of the packet by the source host, as illustrated in Fig. 4.

A Brief History of the Internet

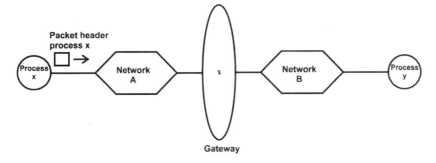

Fig. 3 Internetworking by Gateway (adapted from [3])

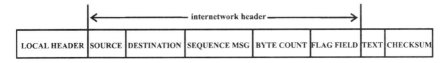

Fig. 4 Internetwork Header (adapted from [3])

The internetwork header contained the standardized source and destination addresses. The next two fields in the header provided a sequence number and a byte count used to properly sequence the packets upon delivery to the destination and also enabling the gateways to detect fault conditions. The flag field was used to convey specific control information. The remainder of the packet contained the payload (text) and a trailing check sum used for end-to-end software verification. The gateway does not modify the information, only forwarded the header check sum along the path.

2.2 TCP

The TCP protocol specified by Cerf and Kahn [3] had the function of promoting the transmission and acceptance of messages of processes that wanted to communicate. To implement this function, TCP first broke the process messages into segments according to a maximum transmission size. This action was called fragmentation and was done in such a way that the destination process was able to reassemble the fragmented segments. On the transmission side, the TCP multiplexed together segments from different processes and produced packets for delivery to the packet switches. On the reception side, the TCP accepted the packets sequence from the packet switches, demultiplexed and reassembled the segments to the destination processes.

This system introduced the notion of ports and TCP address. A port was used to designate a message stream associated with a process. A TCP address was used to routing and delivery packets from diverse processes to the suitable destination host. The original TCP address format is shown in Fig. 5.

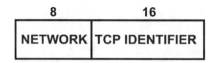

Fig. 5 TCP Address Format (adapted from [3])

The use of 8 bits for network identification (ID) allowed up to 256 distinct networks. At that time, this address field seemed enough for the future. The TCP identifier field permitted up to 65 536 distinct TCP be addressed. As each packet passed through the gateway, it observed the destination network ID to determine the packet route. If the destination network was connected to the gateway, the lower 16 bits of the TCP address were used to produce a local TCP address in the destination network. On the other hand, if the destination network was not connected to the gateway, the upper 8 bits was used to select the next gateway.

In order to send a TCP message, a process settled the information to be transmitted in its own address space, inserted network/host/port addresses of the transmitter and receiver in a transmit control block (TCB), and transmitted it. At the receiving side, the TCP examined the source and destination port addresses and decided whether accepted or reject the request. If the request was rejected, it merely transmitted a release indicating that the destination port address was unknown or inaccessible. On the other hand, if the request was accepted, the sending and receiving ports were associated and the connection was established. After it, TCP started the transmission of the packets and waited for the acknowledgements carried in the reverse direction of the communication. If no acknowledgement for a particular packet was received, the TCP retransmitted the packet.

Aftertime, a window strategy to flow control of sent and received packets also was proposed by Cerf and Kahn [3], as shown in Fig. 6.

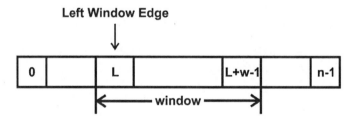

Fig. 6 Window Strategy (adapted from [3])

Supposing that the sequence number field in the internetwork header permits packet sequence numbers to range from 0 to n − 1, the sender could not transmit more than w bytes without receiving an acknowledgment. The w bytes were named a window (see Fig. 6). On timeout, the sender retransmits the unacknowledged bytes. Once acknowledgment was received, the sender's left window edge advanced over the acknowledged.

A Brief History of the Internet 7

After the development of fundamental characteristics of TCP, the next challenge of DARPA was running TCP on multiple hardware platforms and making experiments to determine optimal parameters for the protocol. In 1977, the ARPA research program included important players in this development such as: BBN, DCEC, ISI, MIT, SRI, UCLA and some prototypes of TCP/IP were implemented.

2.3 Ethernet

At the same time, the development of the first concepts of new computer networking technology for local area networks (LANs) named Ethernet. This technology has been widely used under TCP/IP and has grown importance to encompass new technologies.

The Ethernet idea began on May 22, 1973, when Bob Metcalfe (then at the Xerox Palo Alto Research Center, PARC, in California) wrote a memo describing the Ethernet network system he had invented for interconnecting advanced computer workstations, making it possible to send data to one another and to high-speed laser printers (see Fig. 7). The seminal article: "Ethernet: Distributed Packet Switching for Local Computer Networks" was published by Robert M. Metcalfe and David R. Boggs in [6].

Robert Metcalfe got the idea for the Ethernet protocol when he read a 1970 computer conference paper by Norman Abramson of the University of Hawaii about the packet radio system called ALOHANET linking the Hawaiian Islands. At the end of 1972, the ALOHANET was connected to ARPANET by satellite given a pass to the development of the Internet.

Each node in ALOHANET sent out its messages in streams of separate packets of information. If it did not get an acknowledgment back for some packets because

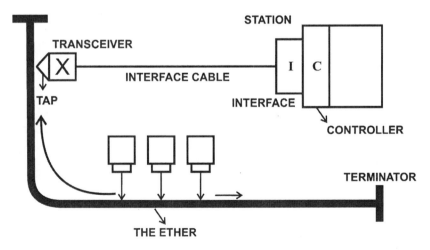

Fig. 7 Robert Metcalfe picture and his famous Ethernet first drawing (adapted from [4])

two radios were broadcasting at the same time, then the missing packets were considered "lost in the ether". The word *ether* was used to denote the propagation medium that could be used by any type of machine, in analogy to the material believed by the physicists to fill in the free space enabling the electromagnetic propagation.

When a packet was lost in the ether, the node would re-broadcast them after waiting a random interval of time. Because of this randomness, problems with collisions were quickly resolved except under very high traffic loads. On average, the network rarely had to retry more than once or twice to get all the packets to the destination, which was more efficient than trying to implement a complex coordination system to prevent collisions in the first place. The original 10 Mbps Ethernet standard was first published in the next decade by the DEC-Intel-Xerox (DIX) vendor consortium.

3 Decade of the 80´s

After testing three increasingly better versions: TCPv1, TCPv2, a split into TCPv3 and IPv3, finally in 1981, TCP (Transmission Control protocol) v4 and IP (Internet Protocol) v4, posted in RFC 791 [7] and RFC 793 [9], respectively, became stable. This version is still in use on the Internet today.

In 1982, an Internet Gateway, to route internet packets based on TCPv4/IPv4, developed by BBN, was standardized in RFC 823 [5]. TCPv4/IPv4 became a standard for DARPA and, in January, 1983, the ARPANET protocol switched from NCP to TCP/IP. This date is considered the date of the birth of the Internet [11]. In 1985, Dan Lynch and the IAB (Internet Architecture Board) realized a workshop for the computer industry to become TCP/IP a commercial standard and promote the development of networking products.

3.1 Internet Protocol

The IPv4 implements two basic functions: fragmentation and addressing. Fragmentation is used to break the packets in small pieces to transmit through "small packet" networks.

The addressing is used to forward Internet packets toward their destinations. The Internet protocol treats each Internet packet as an independent entity. There are no connections or logical circuit establishment. So, the Internet protocol does not provide a reliable communication facility, only hop-by-hop forwarding of packets. There is no error control for the information, only a header checksum and errors detected in the header are reported via the Internet Control Message Protocol (ICMP) [8].

The Internet transmission occurs when an application program via transport protocols sends a request on its local router (gateway) to send data as a packet thought the Internet (see Fig. 8). The Internet router prepares the packet header and attaches it to the data. The router determines a local network address and sends the packet to the local network interface. The local network interface creates

A Brief History of the Internet 9

a local network header, attaches it to the packet and sends it to the local network. The packet is forward hop-by-hop through the network until the local network where the destination host is located. At each hop, the router examines the header and determines the next hop based on the destination address. At the destination router, the packet is sent to the destination host, via transport protocol socket to the application.

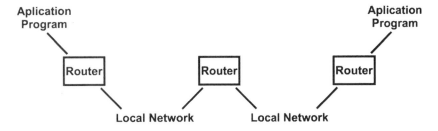

Fig. 8 Internet Forwarding Packets (adapted from [3])

3.2 Ethernet Protocol

The first Ethernet standard was entitled "The Ethernet, A Local Area Network: Data Link Layer and Physical Layer Specifications" and was published in 1980 by the DIX vendor consortium. It contained the specifications of both the operation of Ethernet and the single media system based on thick coaxial cable.

Ethernet is by definition a broadcast protocol where any signal can be received by all hosts. The packets from the network layer are transmitted over an Ethernet by encapsulating them in a frame format as shown in Fig. 9.

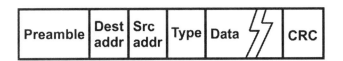

Fig. 9 Ethernet Frame

The fields of this frame are described as follow:

Preamble: is a sequence of 8 bytes, each set to "10101010" and used to synchronize receiver before actual data is sent;
Addresses
- 48-bit unicast address assigned to each adapter, named MAC (Medium Access Control) Address
- Broadcast address: all bits set to 1
- Multicast: first bit is set to 1

Type field: is used to determine which higher level protocol the frame should be delivered to
Body: contains up to 1500 bytes of data

When the Ethernet standard was published, a new effort led by the Institute of Electrical and Electronics Engineers (IEEE) to develop open network standards was also getting underway. The IEEE standard was created under the direction of the IEEE Local and Metropolitan Networks (LAN/MAN) Standards Committee, which identifies all the standards it develops with the number 802. There have been a number of networking standards published in the 802 branch of the IEEE, including the 802.3 Ethernet and 802.5 Token Ring standards. The IEEE 802.3 committee took up the network system described in the original DIX standard and used it as the basis for an IEEE standard. The IEEE standard was first published in 1985 with the title IEEE 802.3 Carrier Sense Multiple Access with Collision Detection (CSMA/CD). Ethernet uses CSMA/CD to listen the line before sending data:

- If the line is idle (no carrier sensed), it sends packet immediately;
- If line is busy (carrier sensed), it wait until idle and transmit packet immediately;
- If collision is detected, it stops sending and try again later.

After the publication of the original IEEE 802.3 standard for thick Ethernet, the next development in Ethernet media was the thin coaxial Ethernet variety, inspired by technology first marketed by the 3Com Corporation. When the IEEE 802.3 committee standardized the thin Ethernet technology, they gave it the shorthand identifier of 10BASE2. Following the development of thin coaxial Ethernet came several new media varieties, including the twisted-pair and fiber optic varieties for the 10 Mbps system. Next, the 100 Mbps Fast Ethernet system was developed, which also included several varieties of twisted-pair and fiber optic media systems. Most recently, the Gigabit Ethernet and 10 Gigabit Ethernet systems were developed and 100 Gigabit Ethernet is in development. These systems were all developed as supplements to the IEEE Ethernet standard.

3.3 Evolution of Internet

In 1985, the National Science Foundation (NSF) launched a network to connect academic researchers to supercomputer centers to provide very high-speed computing resources for the research community. This network was named NSFNET and one of its project design premises was to use ARPANET's TCP/IP protocol. In 1986, the NSFNET was connected to ARPANET and these backbones forming what today is known as Internet. At the end of this decade, NSFNET became de facto the backbone of the Internet and the ARPANET was ended (Stewart 2000).

Also in this period, the World Wide Web (WWW) system was created by Tim Berners-Lee [1] to run in the Internet and provide graphical user interfaces and hypertext links between different addresses.

In 1991, the Internet became commercially exploited and new backbones were built to offer services of communications. This fact became Internet completely decentralized, without a central coordination, difficult architectural changes. In 1995, the NSFNET was officially dissolved, although, retained a core research network called the Very High Speed Backbone Network Service (vBNS), which formed the basis for the Internet2 project [10].

Since 1995, the Internet continues growing; more and more people use it to be connected, find information, create business, and share information. The Internet is now an essential part of our lives.

References

1. Berners-Lee: Information Management: A Proposal, CERN (1989), http://www.w3.org/History/1989/proposal.html (accessed March 2010)
2. Bolt, Beranek, Newman: Report No. 1822: Specification for the Interconnection of a host and an IMP (1976)
3. Cerf, V., Kahn, R.: A Protocol for Packet Network Intercommunication. IEEE Transactions on Communication 22(5) (1974)
4. Cocker, S., Carr, S., Cerf, V.: RFC 33 New Host-Host Protocol (1970)
5. Hinden, R., Shelzer, A.: RFC 823 DARPA Internet gateway (1982)
6. Metcalfe, R., Boggs, D.: Ethernet: Distributed Packet Switching for Local Computer Networks. Communications of the ACM 19(5), 395–404 (1976), http://www.acm.org/classics/apr96/ (accessed March 2010)
7. Postel, J.: RFC 791 Internet Protocol (1981)
8. Postel, J.: RFC 792 Internet Control Message Protocol (1981)
9. Postel, J.: RFC 793 Transmission Control Protocol (1981)
10. Stewart, B.: Living Internet (2000), http://www.livinginternet.com/i/i.htm
11. Wladrop, M.: Darpa and the Internet Revolution. DARPA 78-85 (2008), http://www.darpa.mil/Docs/Internet_Development_200807180909255.pdf (accessed March 2010)

Principles of Internet Architecture

Tania Regina Tronco

CPqD Foundation, Rodovia Campinas Mogi-Mirim, km 118,5,
Campinas – São Paulo, CEP 13096-902, Brazil
tania@cpqd.com.br

Abstract. This chapter reviews the original Internet architecture with special focus on its original requirements and principles. The motivation behind this "flash-back" is that the Internet architecture is evolving fast and understanding its original design principles provides a context to study the new architectural design challenges necessary nowadays.

1 Introduction

The design of the original Internet protocols, namely TCP/IP suite (Transmission Control Protocol/Internet Protocol), was guided by a set of design principles but only in 1988, the Internet architecture was formalized by David Clark [4]. Clark catalogued a set of goals and features of the Internet such as:

- The Internet architecture should provide a standard protocol to interconnect ARPANET with other sorts of networks using a packet switching technique. The packet switching technology was chosen because enabled the integration with other existing packet network technologies such as packet radio networks, broadcast satellite networks and local area networks.
- The second goal was survivability (robustness), as it was designed to operate during the Cold War. The Internet should maintain the communication service even during temporary disruptive events.
- The third goal of Internet architecture was to support diverse types of service at the transport level. The TCP provides a reliable delivery of data and it is suitable for applications such file transfer, not concerned with delays. So, new protocols at the transport layer were developed such as UDP (User Datagram Protocol) to provide basic datagram services to the Internet [4], not reliable but with low delay requirements.

The goals described above were considered priorities in a military context of that time, but there were other goals in the Internet architecture, considered less important, as shown in the following list:

- The Internet should allow distributed management of its resources;
- The Internet architecture should be cost effective;
- The Internet architecture should permit easy host attachment;
- The resources used in the Internet architecture should be accountable.

The goal of distributed management is concerning with the Internet support for multiples administrative regions (domains) managed by different management centers. The gateways should exchange routing tables to perform routing computation in a distributed way avoiding interruptions in the network due a single point of failure.

As mentioned in [4], this set of goals are in order of importance, a different network architecture would result if the order were changed.

The goals listed above carry out a series of choices based on which it is possible to derivate some the important principles and the basic mechanisms that guided the original Internet architecture, as follows:

- Connectionless packet-forwarding service;
- Transparency;
- Survivability;
- Different types of quality of service;
- Globally addresses;
- Layered Protocol Stack;
- Distributed Management;
- No mobility;
- No security.

In the following sections, these principles will be detailed explaining the resultant features of Internet architecture.

2 Connectionless Packet-Forwarding Service

One key characteristic of the Internet is the use of variable-length packet as the basic entity. The Internet packet is composed by a header and an information field as shown in Fig. 1.

The header's fields are described as follow:

Version (4 bits): indicate the standard of the Internet header, e.g., version 4.

Internet Header Length - IHL (4 bits): is the length of the Internet header in 32 bit words, pointing the beginning of the data. (Minimum value is 5).

Type of Service (8 bits): is used to specify the type of treatment of the packet during its transmission through the Internet, providing an indication of the quality of service desired.

Principles of Internet Architecture 15

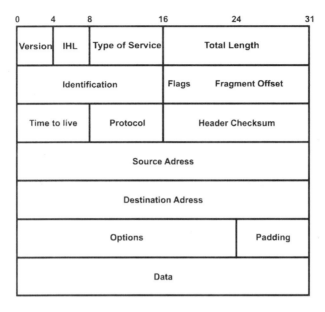

Fig. 1 IP Header

Total Length (16 bits) is the length of the packet including the Internet header and information. The size of this field allows the length of the packet to be up to 65,535 bytes.

Identification (16 bits): value assigned by the sender to assembling the fragments of the packet.

Flags (3 bits):

Bit 0: reserved, must be zero
Bit 1: (DF) 0 = May Fragment, 1 = Don't Fragment.
Bit 2: (MF) 0 = Last Fragment, 1 = More Fragments.

Fragment Offset (13 bits): indicate where in the packet this fragment belongs. The fragment offset is measured in units of 8 bytes (64 bits). The first fragment has offset equal to zero.

Time To Live (8 bits): indicate the maximum time that the packet is allowed to remain in the network. If this field contains the value zero, then the datagram must be destroyed. TTL is reduced by one on every hop.

Protocol (8 bits): indicate the next level protocol used in the information field of the Internet packet.

Header Checksum (16 bits): is a checksum of the header and is a verification that the information used in processing Internet packet has been transmitted correctly. It is recomputed and verified at each point where the Internet header is processed.

Source Address (32 bits): indicate the address of the sender.

Destination Address (32 bits): indicate the address of the receiver

Options: provide control functions needed or useful in some situations such as timestamps, security and special routing.

Padding: used to ensure that the Internet header ends on a 32-bit boundary. The padding bits are set to zero.

In the Internet operation, the IP addresses are used as directive for forwarding the packets hop-by-hop across the underlying network. In contrast to the virtual circuit, which usually implies a connection establishment, the Internet is a connectionless infrastructure that multiplexing the packets and forward them as an independent way unrelated to any other internet packet. Each router in the path is responsible for read the IP address in the header of each packet, search the best output interface, based on the destination address, and forward the packet trough this interface, as shown in Fig. 2.

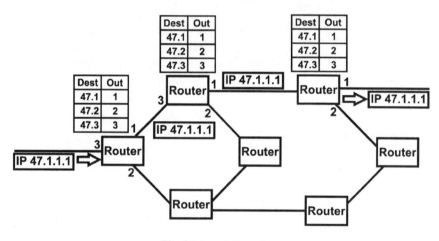

Fig. 2 Internet Operation.

The selection of a path for transmission is called routing.

The IP layer also provides datagram fragmentation and reassembly, so that a datagram originally transmitted as a single unit will arrive at its final destination broken into several fragments packets according with a maximum transmission size (MTU = Maximum Transmission Unit), as shown in Fig.3. The IP layer at the receiving host must assemble these fragments to reconstitute the original datagram.

As a consequence, the IP layer provides a datagram service (connectionless) without guaranteed delivery ("best-effort"). The packets may be dropped, duplicated, or reordered during the delivery.

Principles of Internet Architecture

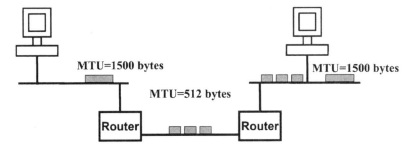

Fig. 3 Packet Fragmentation.

If the IP layer had already be designed for the connection oriented service, it would not be flexible enough to also support datagrama services. As result, the Internet architecture is open to support diverse transport protocols such as TCP and UDP (User Datagrama Protocol) with diverse types of services and different constraints such as: reliability, delay, throughput, etc. This goal was the reason of the split of the TCP in two layers, TCP and IP [4].

3 Transparency and Simplicity

The Internet delivers the user information without modification, providing transparency to users. There is no error control for data, only a header checksum, which provides verification that the information used in processing IP packets has been transmitted correctly. If the header checksum fails, the Internet packet is discarded at the gateway. The elimination of data errors control at Internet layer provided simplicity and generality to the network, reducing costs, making easy network upgrades and facilitating the addition of new applications without changing the network.

These principles were based on the "end-to-end argument" [6] which is a line of reasoning against application-level functions being placed at network layer. Saltzer suggests that these functions can be correctly implemented only with the knowledge and help of the application at the end points of the communication. Moreover, bit error recovery, duplicate message suppression and delivery acknowledgement placed at network layer may be redundant or of little value when an "end-to-end check and retry" were already implemented at application-level.

So, the Internet protocol does not provide a reliable communication facility, only a "best effort" service, packets may be dropped, duplicated, or reordered at network layer, there aren't retransmissions neither flow control at this layer. This way of operation is sometimes called "dumb network" but is this feature that enables the creation of new applications at the edge of the net ("smart" end systems), increasing the innovative potential of the Internet [1].

4 Survivability

The Internet connectionless service provides survivability once that when a failure occurs, there is not connection broken at network layer and, at the transport layer; TCP can recover the information, sending again the packets lost and reordering the packets at the end points. So, TCP storages the state information such as number of packets transmitted and acknowledged at the endpoints of the net, instead of Internet store this information at the intermediate nodes. In [4], this approach to reliability was named "fate-sharing". This means that is acceptable to lose the state information for an entity if and only if the entity itself is lost (see Fig. 4), i.e. it is acceptable to lose TCP state if one endpoint crashes but not acceptable to lose it if an intermediate router fail.

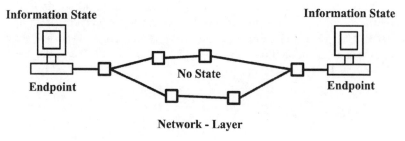

Fig. 4 Fate-Sharing

5 Types of Quality of Service

The Internet protocol uses the "Type of Service" field (8 bits) in the packet header to indicate the quality to be provided by the network. This indication is used only to select parameters in the gateway. The types of Quality of Service (QoS) available are:

Bits 0-2:	Precedence: high precedence traffic is more important than other traffic.
Bit 3:	0 = Normal Delay, 1 = Low Delay.
Bits 4:	0 = Normal Throughput, 1 = High Throughput.
Bits 5:	0 = Normal Reliability, 1 = High Reliability.
Bits 6-7:	Reserved for Future Use.

6 Globally Fixed Addresses

In the original Internet architecture, the packets are passed from one subnet to another, until they reach their destination using fixed-length global IP address. These IP addresses are represented by four decimal numbers, separated by dots between them (x.y.z.w) and divided into classes, being each class consists of 32-bit words, containing the network identification and the host belonging to it, as follows:

Principles of Internet Architecture

- Class A: 1.0.0.0 to 127.255.255.255
- Class B: 128.0.0.0 to 191.255.255.255
- Class C: 192.0.0.0 to 223.255.255.255
- Class D: 224.0.0.0 to 239.255.255.255
- Class E: 240.0.0.0 to 247.255.255.255

The division in classes facilitates the reading of the address of the sub-networks by gateways. The simple test of initial bits identifies the class used, enabling the aggregation of information and easy next-hop look-ups at routers.

IP addresses are placed in a predetermined field in the header of IP packets, indicating the source address (source address) and destination (destination address), as shown in Fig 1.

7 Layered Protocol Stack

The TCP/IP architecture uses four layers [2] to decompose the protocol into functional modules: application layer, transport layer, Internet layer and link layer, as shown Fig.5.

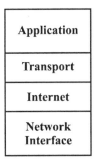

Fig. 5 TCP/IP Layers.

The application layer threats the syntax and semantics of information and provides applications to users, such as:

- Telnet (remote login)
- FTP (file transfer)
- SMTP (electronic mail delivery)

The transport layer provides end-to-end communication services for applications. There are two basic transport layer protocols:

- Transmission Control Protocol (TCP)
- User Datagram Protocol (UDP)

TCP is a reliable connection-oriented transport service that provides end-to-end reliability, re-sequencing, and flow control. UDP is a connectionless ("datagram") transport service.

The Internet Layer provides a connectionless or datagram internetwork service, providing no end-to-end delivery guarantees. Thus, IP packets may arrive at the destination host damaged, duplicated, out of order, or not at all.

The link layer is used as media-access protocol and there is a wide variety of link layer protocols, corresponding to the many different types of access networks. The TCP/IP architecture has an "hourglass" shape, in which a wide variety of applications and transport protocols are supported by a single, "narrow" protocol called IP, which in turn rests upon a wide variety of network and link protocols, as shown in Fig.6.

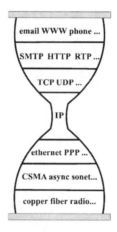

Fig. 6 Hourglass Model for the Internet (adapted from [5])

This hourglass design has provided enormous flexibility to the Internet assuming least common network functionality (IP) to maximize the number of usable networks at link layer, isolating end-to-end protocols from network and accommodating new applications. "Anything over IP" and "IP over anything" has allowed innovation both above and below the IP layer.

8 Distributed Management

Internet supports multiple autonomous systems (AS), independently managed, by exchanging routing information in a distributed way. The generation of routing tables is, in general, automatic and uses a distributed routing algorithm that performs calculations based on the metrics of the links.

There are two standard distributed routing algorithms:

- **Distance Vector**, which builds the routing table considering the smallest distance to each destination gateway. The distance is measured in number of hops, or number of routers to reach the gateway. The update of the table is done via communication between neighbors, where each router sends distance vectors to its neighbors containing its shortest distance to each destination router. Upon receiving the distance vectors from a neighbor, the router updates its routing table. The Bellman-Ford algorithm [3] is used to compute these shortest paths. To reduce the chance of forming loops, a split horizon with poison reverse technique is implemented. This technique consists in sending information of distance infinity of a destination to a neighbor if it is the next hop to that destination. This mechanism also accelerates the routing convergence.

- **Link State**, which each router announces every other router its distance to its neighbors. The basic idea is distribute to all routers the topology of the network and each router independently computes optimal paths and has global view of the network. Djkstra's algorithm [3] is used to compute shortest path to each destination and different metrics, called costs, may be associated to the links. The cost can be number of hops, delay, speed, bandwidth or a weight arbitrarily assigned by the administrator.

*At the beginning operational of the Internet, at intra-domains, the rout*ing algorithm used was distance vector and the routing protocol was RIP (Routing Information Protocol). Since 1988, a link-state algorithm named OSPF (Open Shortest Path First) has been adopted intra-domains because of its fast convergence compared to RIP considering the growth of the size of the network. On the other hand, to exchange routing information between autonomous systems (inter-domains), the BGP (Border Gateway Protocol) has been used.

BGP is a path vector routing protocol, a variant of the distance vector that lists the explicit full path of an AS to reach another AS. It is used in inter-domains because of policy control characteristic where, policy decisions may be enforced differently at each AS. These policy decisions can be related to political, security, or economic considerations, becoming BGP very complex [7].

9 No Mobility

The other feature of the Internet architecture is that addresses applied to physical interfaces of the routers and hosts are used for two functionalities: routing and naming of the interface; if the router or host changes its physical location, the IP address changes. As shown in Fig.7 and Fig.8, when the host with IP address 223.1.1.2 changes its location to another subnet, its address changes to 223.1.3.3.

Hence, in the original Internet architecture, there is no distinction between a node identifier and the network attachment point (location), leading to a "semantic overload" of the IP addresses. As a result, the original architecture considers only static network attachments, no mobility.

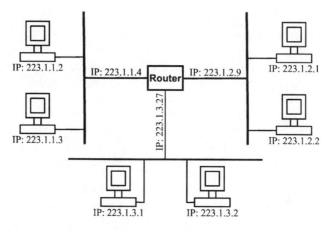

Fig. 7 Host Mobility

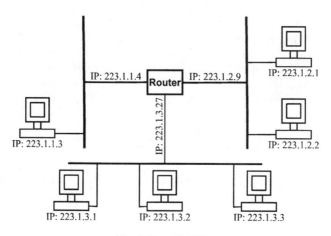

Fig. 8 Host Mobility

10 No Security

The last original Internet architecture principle mentioned here is that security is considered as a host problem i.e. no security should be implemented at network layer). This principle is based on the "end-to-end argument" [6] that suggest that security using encryption must be implemented at the application layer.

11 Discussion

In this chapter, we recalled the main Internet architectural principles of the original Internet. The original design concepts present some very ideas in terms of simplicity, survivability and distributed management but also present some gaps in terms of mobility, security and addressing space. A lot of engineering studies are required to address these problems and provide a more powerful architecture to Internet. In the next chapter, the evolution of the Internet to now will be discussed.

References

1. Blumental, M., Clark, D.: Rethinking the design of the Internet: The end to end argument vs. the brave new world. ACM Transcations on Internet Technology 1, 70–109 (2001)
2. Braden, R.: RFC 1122 Requirements for Internet Hosts - Communication Layers (1989)
3. Chuah, C.: Computer Networks Packet Switching Networks Routing in Packet Networks Shortest Path Algorithms: Dijkstra's & Bellman-Ford Reading: ch. 7.4 & 7.5 (2004), http://www.ece.ucdavis.edu/~chuah/classes/EEC173A/lectures/L10_routing.pdf (accessed March 2010)
4. Clark, D.: The Design Philosophy of the DARPA Internet Protocols. In: SIGCOMM 1988, Computer Communication Review, vol. 18(4), pp. 106–114 (August 1988)
5. Deering, S.: Watching the Waist of the Protocol Hourglass, IETF 51, London (August 2001), http://www.ietf.org/proceedings/01aug/slides/plenary-1/tsld001.htm (accessed March 2010)
6. Saltzer, J., Reed, D., Clark, D.: End-to-End Arguments in System Design. In: Second International Conference on Distributed Computing Systems, April 1981, pp. 509–512 (1981)
7. Sherwood, R.: BGP Tutorial, Stanford CS144 (October 2009)

Evolution of Internet Architecture

Tania Regina Tronco

CPqD Foundation, Rodovia Campinas Mogi-Mirim, km 118,5,
Campinas – São Paulo, CEP 13096-902, Brazil
tania@cpqd.com.br

Abstract. The Internet architecture has evolved along several dimensions since its original conception and still is in continuous evolution. Extensions to the IP original architecture have been incrementally proposed since the seventies in order to have specific requirements. This chapter surveys the main changes following a chronological order.

1 Introduction

Ad-hoc extensions to the original Internet architecture have been developed to solve individual issues, such as addressing space exhaustion, lack of Quality of Service (QoS), lack of security, performance increasing, mobility, etc. These extensions will be briefly described here as follows: Classless Inter-Domain Routing (Section 2), Network Address Translation (Section 3), Dynamic Host Control Protocol (Section 4), Domain Name System (Section 5), Firewalls (Section 6), IP version 6 (Section 7), QoS (Section 8), Multiprotocol Label Switching (Section 9), New Ethernet (Section 10), Virtual Private Networks (Section 11), Caching, replication and Content Delivery Networks (Section 12), Peer-to-Peer networks (Section 13), IP Mobility and Multi-homing (Section 14). Finally, Section 15 discusses some of the inconsistencies between the extensions and the principles of the Internet architecture.

2 Classless Inter-Domain Routing (CIDR)

Originally, the addressing space was divided in three fixed classes, being each class consisting of a 32-bit word, containing a network identifier (*netid*) and the host identifier (*hostid*), as shown in Fig.1.

Class A accommodates up to 128 networks and up to 16 million hosts each;
Class B accommodates up to 16,382 networks with up to 64K hosts and
Class C accommodates up to 2million networks and up to 254 hosts.

This division in fixed classes aimed simplifying sub-network address handling by routers, which could perform a simple test on initial bits to identify the class used,

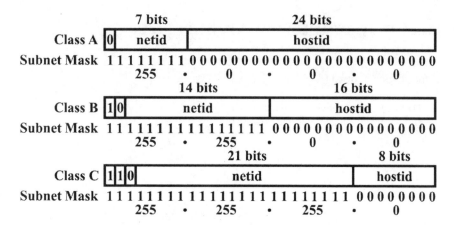

Fig. 1 IP Address Fixed Classes.

increasing the efficiency of the network service. As shown in Fig.1, a subnet mask determines where the network address ends and the host address begins. The division between networks and hosts is closely tied to the separation of intra-domain routing and the inter-domain routing.

As the Internet evolved, some serious scaling problems came out, such as: the exhaustion of IP addressing space and the overgrowth of routing tables. The address division in classes with fixed length for network and host identification soon became inappropriate because it was not flexible to adapt to real situation in terms of number of hosts and networks each one. Then, in RFC 1518 [45] and RFC 1519 [8], a technique of Variable-Length Subnet Mask (VLSM) where the network/host divide can occur at any bit boundary in the address was proposed. The IP address becomes represented by a.b.c.d/x, where x define the size of the network portion in address and the length of the subnet mask.

As an example, the IP address 136.122.10.192/28 represents 16 possible addresses (32 bits – 28 bits = 4 bits => 2^4 = 16) ranging from 136.122.10.192 until 136.122.10.207.

Due to normal class division being ignored, this system was called Classless Inter-Domain Routing (CIDR).

Another benefit of CIDR is the possibility of aggregation of routing prefixes. For example, sixteen networks /24 contiguous can be aggregated as a single route /20 (if the first 20 bits of network address match). Two /20 contiguous can be aggregated into a /19, and so on. The CIDR establishes a hierarchical sub-allocation of addresses allowing a significant reduction of the routing table size. Some changes and adaptations were necessary in the routing protocols to support the concept of route aggregation and VLSM.

3 NAT (Network Address Translation)

As the Internet was growing beyond anyone's expectations, a unique global address space would become exhausted, if additional measures had not been taken.

In fact, in IPv4, the number of routable IP addresses is not sufficient if all machines were connected to the Internet. Then, Internet Assigned Numbers Authority (IANA) reserved the following blocks of the IP address space for private networks, standardized in RFC 1918 [46] as follows:

10.0.0.0   - 10.255.255.255 (10/8 prefix)
172.16.0.0 - 172.31.255.255 (172.16/12 prefix)
192.168.0.0 - 192.168.255.255 (192.168/16 prefix)

Additionally, a mechanism of Network Address Translation (NAT) was developed in RFC 3022 [51] to use a globally routable IP address to connect machines using private address to Internet. This mechanism performs a translation of the private LAN IP addresses (not routable) into a routable Internet IP address. A gateway is used to implement this functionality at the LAN boundary, as shown in Fig. 2.

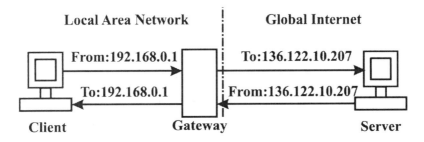

Fig. 2 Network Address Translation

In Fig.2, the gateway associates a private IP address (e.g. 192.168.0.1) to a public routable IP address (e.g. 136.122.10.207) on the Internet and vice-versa, enabling the client machine to be connected to the public Internet.

NAT benefits beyond address expansion are:

- Security: the local area networks internal address numbering is hidden from public Internet;
- Flexibility: when the public routable IP addresses changes, the gateway at the edge of the network is the only device that requires a reconfiguration;
- Eliminate immediate renumbering efforts: NAT allows the existing address scheme to remain at local area network even public Internet changes.

Recently, a Carrier-Grade NAT (CGN) has been proposed to share IPv4 addresses among end-users and the public Internet, preventing global IPv4 address consumption [26]. CGN may also share address between IPv4 end-users and IPv6 end-users, as shown in Fig.3. In order to provide this functionality, a Network Address Translator - Protocol Translator (NAT-PT) mechanisms was necessary [53].

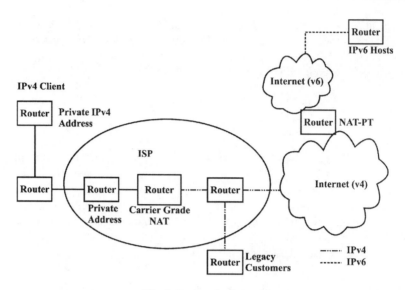

Fig. 3 Carrier-Grade NAT

4 DHCP (Dynamic Host Configuration Protocol)

DHCP mechanism was developed to minimize renumbering efforts providing capabilities to automatic allocation of IP addresses. This mechanism assigns an IP address to a host (client) for a limited period of time and reuses it after the machine is turned off. Supposing that not every machine on the network is turned on at the same time, it's possible to dimension more users in the network than the number of IP addresses assigned to the DHCP server. DHCP detailed specification is in [6].

5 DNS (Domain Name System)

The DNS allows the use of the names (domains) instead of using IPs address to access Internet sites. It consists of a set of large distributed databases on root servers around the world indicating what IP address is associated with a name and vice-versa. Then, DNS provides a Domain Name-to-IP address translation and a IP address-to-Domain name translation, allowing the location of hosts in a domain and a domain name associated to host identification, respectively. Fig.4 illustrates an example of domain name-to-IP address translation.

In Fig.4, DNS client wants to get the IP address for www.domainname.com. It sends a request to DNS server that returns 192.168.1.12 as IP address for the required domain.

DNS is a key component in the Internet architecture but as Web pages become more complex, referencing different domains, as well the number of users of the Internet grows, DNS look-ups can become a significant bottleneck in the browsing

Fig. 4 Example of Domain name-to-IP Translation

experience. Moreover, DNS is always subject to attacks and manipulation to provide load balancing and CDN (Content Delivery Networks) tricks. CDN tricks are used to optimize the experience for the end subscriber. One example of such trick is which the same hosts will get different IP addresses for a domain, depending upon the origin of the look-up, i.e. a subscriber in a particular city will get a different IP address for a domain than a subscriber in another city.

6 Firewalls

Firewall systems had been introduced into Internet to protect the network from attacks and to restrict the access to the network. There are many types of firewalls such as packet filtering, proxy filtering, etc. In packet filtering, each data packet passes through the filters for proper analysis based on firewall administrator rules. Proxy filtering treats the flow of packets as one application and tests them before delivering to the receiver.

7 IPv6

IPv6 was designed to be the successor to IP version 4 (IPv4). His creation is the result of IETF efforts started in 1993, when projections indicated that the address space would become exhausted [33]. The protocol was specified in the first half of 1995.

The IPv6 specification is in RFC 2460 [5] and the main advantages from IPv6 to IPv4 are:

- Larger addressing capability;
- Multicast and anycast addresses;
- Header simplification;
- Auto-configuration;
- Security.

Next, these characteristics will be explained as well the basic functionalities of IPv6.

7.1 IPv6 Header Format

The IPv6 header format is shown in Fig.5.

Version	Traffic Class	Flow Label	
Payload Length		Next header	Hop Limit
Source Adress			
Destination Adress			

Fig. 5 IPv6 Header

The description of IPv6 header field's is the following:

- Version (4-bits): for Internet Protocol version number = 6.
- Traffic Class (8-bits): used to identify different QoS classes of IPv6 packets.
- Flow Label (20-bits): used to label sequences of packets as non-default quality of service or "real-time" service.
- Payload Length (16-bits): indicate the length of the IPv6 payload.
- Next Header (8-bits): used to specify the type of information immediately following the IPv6 header (other protocols' header).
- Hop Limit (8-bits): integer number decremented by 1 by each node that forwards the packet. The packet is discarded if Hop Limit value is zero.
- Source Address: 128-bit address of the originator of the packet.
- Destination Address: 128-bit address of the receiver of the packet.

The IPv6 header has fewer fields compared to IPv4 header and the checksum has been removed. These simplifications reduce the processing and increase the performance of the router, following the end-to-end principle of the original Internet architecture.

7.2 IPv6 Address Format

With the extension of the size of the IP address field from 32 bits to 128 bits, the exhaustion of the addresses IP was solved. The IPv6 address is a very long

Evolution of Internet Architecture

number and difficult to be represented, even in decimal notation. Then, the 128 bits are divided in 8 groups of 16 bits, each one represented by hexadecimal numbers, which ranging from 0000 up to FFFF and separated by colons[11]. An example of an IPv6 address follows:

3FFE:1810:0000:0006:0290:27FF:FE79:7677

An IPv6 unicast address is unique and identifies a single interface. An IPv6 multicast address identifies a set of interfaces and a packet sent to a multicast address is delivered to all interfaces identified by that address. Unicast addresses are distinguished from multicast addresses by the value of the high-order octet of the addresses: a value of FF (11111111) identifies an address as a multicast address. There are no broadcast addresses in IPv6.

There are several forms of unicast address assignment in IPv6 including the aggregatable global unicast address. Similar to CIDR in IPv4, this type of address can be aggregated. The address format is shown in Fig. 6, as follows:

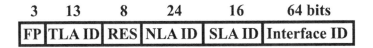

Fig. 6 IPv6 Address Format

Where:

FP	Format Prefix (= 001) for Aggregatable Global Unicast Addresses
TLA ID	Top-Level Aggregation Identifier
RES	Reserved for future use
NLA ID	Next-Level Aggregation Identifier
SLA ID	Site-Level Aggregation Identifier
INTERFACE ID	Interface Identifier and each part of this address

The first 48 bits of the address are used by the IANA for routing between Autonomous Systems (ASes) within Internet. The Site-Level Aggregation (16 bits) is used for routing inside an autonomous system and identifies the destination network. The interface identifier (48 bits) is often the MAC (Medium Access Control) address of the interface at local area network.

When a device is plugged in the network, the local router connected to it sends the IANA prefix and the SLA. Then, the device adds its MAC address and autoconfigure its own IPv6 address. This ability eliminates DHCP server and turns the Internet "plug-and-play".

Moreover, considering the large address space in IPv6, it is possible to decrease the use of firewalls and NAT providing end-to-end connectivity and enabling protocols dedicated to security at the host systems. IPSec (see Section 11.2) is mandatory in IPv6.

Although IPv6 solve the problem of IPv4 addresses exhaustion, global IPv6 deployment has been slower than originally expected, in part because of the large amount of legacy equipments that needed be exchanged, protocols updates and issues related to deploy IPv6 in a mixed IPv4/IPv6 scenario.

8 Quality of Service (QoS)

Extensions to provide QoS to Internet were necessary to meet the growing use of voice and multimedia applications. QoS is indispensable when dealing with real-time applications such as phone calls, teleconference, telemedicine, and so on. One of the techniques to provide QoS is the network resource reservation for different types of applications. In this case, packets will be not discarded if the bandwidth does not exceed pre-defined values by the resource reservation. On the other hand, with "Best Effort" original philosophy of the Internet, packets are dropped randomly in case of network congestion, not assuring the network requirements to the applications.

The IETF defined two models to provide quality of service to IP networks: the Integrated Services (IntServ) [3] and Differentiated Services (DiffServ) [2].

IntServ provides end-to-end quality of services using resource reservation during the establishment of communication. The reservation is done by messages from the signaling protocol RSVP (Resource ReSerVation Protocol)[4]. IntServ also employs admission control, packet classification and scheduling to achieve the desired QoS. The admission control is used to determine whether a data stream can be accepted or not, according to the available bandwidth. The classification of packets considers the ports numbers and protocols types to mark the packets of a specific flow. The scheduler implements algorithms to select which packets will be served first and treat them according to their reservation in the router queue.

Intserv has limitations for scalability due to the signaling protocol RSVP, which requires high storage capacity for monitoring all the flows. This led to establishment of a new simpler and scalable mechanism namely Differentiated Services (DiffServ).

DiffServ [41] is based on the marking bits ToS (Type of Service) in the IP packet header assigning different priority levels for packets. The types of services are classified by categories and use priority criteria to fit the required QoS. DiffServ redefines the layout of the ToS field in IPv4 and the Traffic Class field in IPv6 to DS Field (Differentiated Service Field) as shown in Fig.7.

DSCP Code Point (Differentiated Services) defines different classes (up to 64) for packets forwarding with different treatments by the routers. Packets are classified and marked at network boundary nodes to receive a particular Per-Hop forwarding Behavior (PHB) on the routers along their path. The PHB is indicated by the DSCP field. The most commonly defined PHBs are:

- Default PHB: which is typically best-effort traffic;
- Expedited Forwarding (EF) PHB: dedicated to low-loss, low-latency traffic and
- Assured Forwarding (AF) PHB: which gives assurance of delivery under conditions that the traffic does not exceed some subscribed rate.

Evolution of Internet Architecture

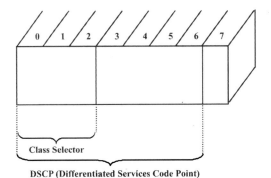

Fig. 7 DS Field

Class Selector PHBs field maintains backward compatibility with the IP Precedence field.

9 MPLS (Multiprotocol Label Switching)

As the Internet grew to become an attractive market, new equipment specifically designed to optimize it were developed.

This approach led to the arrival of solutions to improve the performance of software-based IP routers, simplifying the complex IP over ATM (Asynchronous Transfer Mode) overlay technology. ATM [31], also known as cell relay, was the technology chosen by the ITU (International Telecommunication Union) for the Broadband Integrated Services Digital Network (B-ISDN). Although this technology was meant for the B-ISDN architecture, it had been proven an efficient technique for data transmission, and consequently widely adopted in the backbone networks to transport IP in an overlay model.

At the end of 1996, a number of proprietary solutions were available, which used the characteristics of efficient ATM switching (fast switching via VPI/VCI, called labels), integrating them into the function of IP routing. There were the following solutions:

- IP Switching [40];
- Tag Switching [47];
- Aggregate Route-Based IP Switching (ARIS)[54];
- Cell Switching Router (CSR)[28].

This variety of solutions led the IETF to organize a group to standardize a set of protocols for products that used the process of label switching and had the routing function only at the edges of the network. The resulting network architecture and set of protocols was named Multi-Protocol Label Switching (MPLS) [48]. MPLS is based on a label-switching forwarding algorithm and encapsulates the IP traffic into a new routing header, as shown in Fig.8.

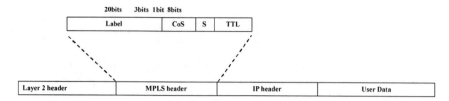

Fig. 8 MPLS Header

In Fig.8, the IP header is encapsulated by a MPLS shim header (or label), before being encapsulated by a layer 2 header. MPLS often referred to as a Layer 2.5 solution.

A label is a fixed-length identifier that is used to forward packets received from a given link. The label value is locally significant to a particular link, and is assigned manually by the network operator or automatically via signaling protocol. Since labels are relatively short, the label of a received packet can be used as an index into a linear array containing the forwarding database. Forwarding database entries indicate the outgoing port and the label to be applied to forwarded frames. Thus, forwarding packets consist of a simple look-up and replacement of the incoming label with the appropriate outgoing label (i.e., label swapping).

The MPLS shim header is also referred to as a label stack, since it may contain multiple entries. Each entry contains a 20-bit label, a 3-bit experimental (EXP) field, a 1-bit End of Stack flag, and an 8 bit Time-To-Live (TTL) field, as shown in Fig.8. The EXP field may be used to identify different traffic classes in support of the DiffServ QoS model. The TTL field is used for loop mitigation in a manner similar to the TTL field carried in IP headers. The End of Stack bit is set to 1 to indicate the last stack entry.

The label-switching is faster because it does not require complex longest-prefix matching on overload IP forward tables. It should be noted that the use of labels is not new in data communications, being used in Frame Relay and ATM.

In MPLS, the connections are unidirectional and named Label Switched Paths (LSPs). There is a variety of mechanisms for establishing LSPs. For example, they may be static, pre-configured by the network operator, or automatically established via protocols such as the RSVP [4].

MPLS protocols are designed primarily for routed IP networks, and are implemented by Label Switch Routers (LSRs). The router where a LSP originates is called the ingress LSR, while the router where a LSP terminates is called the egress LSR. Ingress and egress LSRs are sometimes also referred to as Label Edge Routers (LERs). LERs insert one label into packets transmitted onto a LSP, intermediate LSRs forward packets via label swapping, and egress LERs remove the last label before forwarding packets received from a LSP, as shown in Fig.9.

In Fig.9, at the ingress LER, the IP address of the packet (192.168/16) is examined and, together with information on class of service, it is classified, a label is given to it (5), and it is then routed to the next hop. Next, each LSR uses the incoming MPLS label to forward the packet and to exchange the label (5 to 9 and 9 to 2), and at the egress LER, the label is removed and the packet leaves the MPLS domain with the address original IP (192.168/16).

Evolution of Internet Architecture

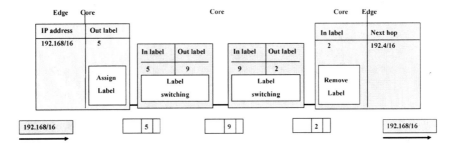

Fig. 9 Label Switching

Once a LSP is established, it can be used to carry IP traffic or to tunnel other types of traffic. The tunnel aspects of LSPs are important in supporting Virtual Private Networks (VPNs), result from the fact that forwarding is based on labels, hidden the original IP address.

Traffic Engineering (TE) is also an important attribute of MPLS optimizing the overall bandwidth utilization in a network by avoiding hyper-aggregation situations, where some portions of the network are overloaded while other portions are under loaded (i.e. traffic engineering can put the traffic where the bandwidth is).

MPLS-TE allow LSPs been established along specified paths (i.e., sequences of specific routers or specific router interfaces), and mapped on Forwarding Equivalence Classes (FECs). A FEC is defined as a group of packets that are forwarded in the same manner (e.g., over the same LSP).

TE capability of MPLS can be performed manually, off-line by a management platform, or dynamically by the MPLS network elements. In off-line case, a management platform may execute algorithms to optimize the overall bandwidth utilization. Then, LERs may be configured with the resulting policies in the form of specific LSP paths and FEC-to-LSP mapping. In online case, routing protocols, such as OSPF, may have its functions expanded to report information about the current bandwidth utilization of each network link. This information may be used to implement a more adaptive form of constraint-based routing, where MPLS routers dynamically calculate explicit paths for LSPs that satisfy configured bandwidth policies (e.g. guarantee a specified minimum bandwidth for traffic passing through a particular egress LSR). Then, the required bandwidth may be reserved along the path when the LSP is established.

MPLS supports both the IntServ and DiffServ models. The IntServ model is supported via RSVP extensions that enable LSPs to be setup in conformance with end-to-end QoS requirements. The DiffServ model is supported mapping different traffic classes to different LSPs via the EXP bits in the MPLS shim header.

In summary, the key attributes of MPLS are:

- Reduces the processing of routers improving efficiency in the delivery of packets;
- Improves IP networks QoS;
- Solves the problem of scale IP/ATM, where IP routers have to be interconnected by a mesh of manually configured ATM connections;

- Facilitate the operation and design of the network using IP developments as a base;
- Separate control (routing) and forwarding functions;
- Operates over any layer 2 technology (from Ethernet to the optical).

The MPLS equipments have already been industrialized and there are a large number of MPLS networks in operation. An agreement on implementation has beeing implemented in the MPLS Forum, allowing interoperability between different vendors equipment.

Since 2005 the ITU also has been adapting MPLS, originally conceived by the IETF, to make a carrier-class networking aligned with the principles of MPLS architecture. This work resulted in the Transport MPLS (T-MPLS), which is a subset of the features of MPLS to provide connection oriented packet switching technology to be used in a transport network. The first three ITU recommendations of the T-MPLS were produced in 2006 as follows:

- Architecture of transport MPLS (T-MPLS) Layer Network [22];
- Interfaces for the Transport MPLS (T-MPLS) Hierarchy [23];
- Characteristics of Multi-Protocol Label Switched (MPLS) equipment functional blocks [24].

In 2008, a Joint Working Team (JWT) was created [1] to find an ITU/IETF agreement and create a MPLS transport profile within the IETF architecture. The objective was extending IETF MPLS with OAM, survivability, network management and control plane protocols to meet transport network requirements as defined by ITU. As a result of this work, a MPLS with Transport Profile (MPLS-TP) was created supporting static and dynamic provisioning, point-to-point, point-to-multipoint, multipoint-to-multipoint traffic, legacy technologies, OAM and reliability, displacing the old T-MPLS.

10 Ethernet Evolution

Ethernet and optical technologies have influenced the Internet backbone technology, especially providing higher speed interfaces such as 40Gbit/s and 100 Gbit/s standards [19].

A lot of technical improvements substantially have been changing the standard Ethernet technology. Ethernet has been used in a number of environments, since LANs, where it was conceived, up to the Wide Area Network (WAN). Originally, Ethernet LANs were based on coaxial cables, with each network device tapping into the cable. This resulted in poor performance and made it difficult to manage. As Ethernet LANs continued to grow, the network devices were linked to a hub using twisted-pair wiring. This solution makes use of only 40% of available capacity as only one host can transmit or receive data at a time for avoid collisions between them and the hosts have to compete for available bandwidth. In 1989, Kalpana Company developed the first Ethernet switch. Ethernet switches can learn which stations are associated with each of its ports, sending the traffic only to the stations which the packet is addressed. Ethernet switches perform their functions using programmable electronic circuits (hardware) instead of software. This provides high-speed packet processing.

Over time, Ethernet switches replaced the hubs and enabled simultaneous switching of data packets between its ports, increasing the performance.

VLAN (Virtual LAN) concept was developed and it is the main mechanism used in switches to segregate the traffic and providing extra security. VLAN is a collection of devices that communicate as if they were on the same broadcast domain. It provides a way of defining the size of Layer 2 broadcast domain and the members belonging to it, independent of their physical location. Moreover, each port of a switch can be configured as attached to distinct VLANs. For instance, one Ethernet port operating at 1 Gigabit/s can be used by several independent VLANs. With this procedure the service price per user can be drastically reduced. Devices within each VLAN can only communicate (at Layer 2) with member devices in the same VLAN. Inter-VLAN communication is done through routing protocols (Layer 3).

There are four types of VLANs: Port-based VLAN, 802.1Q tagged VLAN, MAC-based VLAN and Protocol-based VLAN.

In a port-based, the VLAN is configured as a group of one or more ports on the switch. A port can be member of only one port-based VLAN. A port-based VLAN is also known as an "untagged" VLAN. The disadvantage of this type of VLAN is the need to dedicate multiple ports of a switch to provide one VLAN.

The tagged VLAN [15] solves this problem inserting a marker (called tag) into the Ethernet frame. The tag contains the identification of a specific VLAN, called the VLAN Identifier (VID). With tags, multiple VLANs can share the same Ethernet port.

IEEE 802.1Q has the specification of the Ethernet tagging, as shown in Fig.10. Four octets were included enabling the virtual separation between up to 4096 VLANs (2^{12}) in the VLAN identifier field.

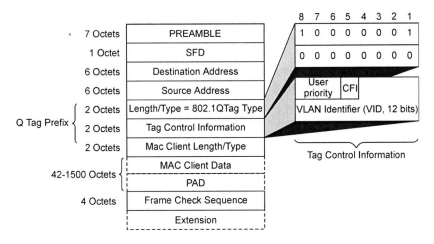

Fig. 10 QTag Ethernet Frame

The user priority field (802.1p) in the tag control information defines up to 8 user priorities.

MAC-based VLANs allows designating a membership by the MAC address of the end stations plugged into the physical ports.

Finally, the protocol-based VLANs enable the administrator to define a packet filter, which is used by the switch as matching criteria to determine if a particular packet belongs to a particular VLAN based on its layer 3 protocols.

Today, the majority of the LAN data traffic starts and ends on Ethernet ports. Deploying Ethernet in the Metropolitan Area Networks (MANs) and WANs is a compelling and commercially proven approach due to cost effectiveness, rapid provisioning on demand, packet-based and easiness of inter-working.

However, there are still some limitations and requirements for a carrier-grade (core-compatible) Ethernet technology such as:

- Lack of end-to-end QoS guarantees;
- Lack of protection mechanisms;
- Lack of in-service OAM (Operation, Administration & Maintenance);
- Lack of scalability;
- Multi-domain and multi-layer resiliency;
- Resource reservation and traffic-engineering.

In order to solve these limitations, some solutions were standardized and others are currently being included in the Ethernet standards.

In terms of protection mechanisms, several options are available, being implemented by industry and proposed in the standards bodies as well:

- Spanning Tree [14]: is a protocol that is used to prevent loops in a redundant network topology. Spanning tree uses the Spanning Tree Algorithm (STA) to calculate the best path through the network. Redundant paths are disabled when the main path is operational and are enabled if the main path fails.
- Rapid Reconfiguration Spanning Tree [17]: implements a faster-convergence algorithm, taking about one second to converge. However, this approach does not meet the restoration times of milliseconds, which may be required by some classes of service.
- Ethernet Automatic Protection Switching [50]: EAPS defined by the IETF (RFC 3619) for ring topologies, is responsible for a loop-free operation and a 50 ms ring recovery.
- Multiple Spanning Tree [18]: allows run multiple spanning trees, each one associated to one VLAN or groups of VLANs.
- Link Aggregation [12] combines the data stream from multiple Ethernet ports into a single high-speed virtual trunk. It provides sub-second failover on trunk groups.

Others possibilities to failure handling in Ethernet networks have been studied such as Simple Network Management Protocol (SNMP), where traps are used to send failures indication to a central manager, and the Bidirectional Forwarding Detection (BFD) protocol [29], which is a hello protocol to rapidly detect failures.

Ethernet has no additional overhead to in-service bit error rate monitoring. The IEEE Ethernet in the First Mile (EFM) working group is specifying Ethernet link OAM using a frame-based approach [13], while ITU are addressing end-to-end in-service OAM for Ethernet in [25].

Considering the scalability requirement, the IEEE 802.1ad defined the so-called "Q-in-Q" or also known as Provider Bridging (PB), to enable a hierarchical VLAN stacking up to 4,096 VLANs for each user and up to 4,096 users. This has increased the scalability of the Ethernet network. Moreover, the IEEE 802.1ah defined the "Provider Backbone Bridges" (PBB) also known as MAC-in-MAC, in which the Ethernet frame is encapsulated in a new MAC address, defined by the service provider, as shown is Fig.11.

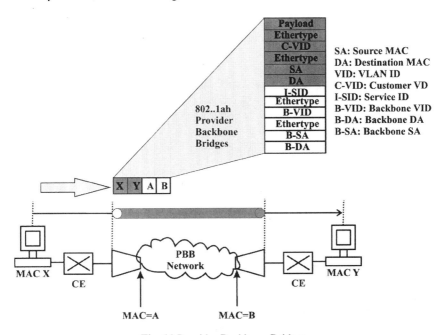

Fig. 11 Provider Backbone Bridges

This solution allows to create tunnels with the new addresses (known only by the provider) avoiding the core switches to process the addresses of all end stations. These addresses are "hidden" by the new MAC, speeding up significantly the processing of information of the backbone. The process MAC-in-MAC inter-operates with Q-in-Q in such a way that both contribute to the scalability and throughput of the Ethernet network. Fig. 12 illustrates this interaction.

In order to enhance the scalability and resource control of the Ethernet networks, the PBB-TE (Provider Backbone Bridges – Traffic Engineering) [16] was also standardized. It is sometimes also referred to as Provider Backbone Transport (PBT) and provides point-to-point tunneling with specific characteristics, such as end-to-end QoS, load balancing, resiliency, etc.

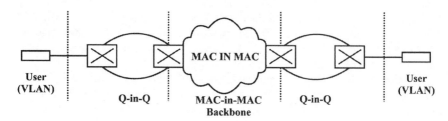

Fig. 12 MAC-in-MAC

Finally, to provide connection oriented in Ethernet networks, the GMPLS (Generalized Multi-Protocol Label Switching) control plane can be used. In [34], GMPLS is defined as an extension of MPLS for a variety of network technologies such as IP, Ethernet, SDH (Synchronous Digital Hierarchy) [21] and OTN (Optical Transport Network) [20]. GMPLS control plane takes care of dynamic provisioning, distributed restoration, network topology and resource discovery. Dynamic provisioning refers to offering bandwidth on demand and end-to-end provisioning. Distributed restoration refers to failure restoration process carried out by the network nodes themselves and not by a centralized network management. Finally, network and resource discovery refers to procedures to automatically discover new ports and new nodes installed or made available in the network.

Therefore, the combined use of GMPLS and PBB-TE contributes to enhance the scalability, end-to-end QoS, resource reservation and resiliency of the Ethernet solution. This set of protocols has been named Ethernet Label Switching (ELS) which establish Ethernet Label Switched Paths (E-LSPs) using the VLAN identifiers (VID) as labels and translating them through the path. This technology maintains the router functionality at the edge of the network and the switching functionality at the core. Moreover, it has the advantage of not needing an additional network layer to provide the connection oriented network service.

Nowadays, there are two groups working on carrier-grade and circuit-oriented solutions: an Ethernet-centric such PBB, PBB-TE and other IP/MPLS-centric. The Ethernet-centric solution is about to build a new network to support Ethernet services in the WAN. The IP/MPLS-centric solution is about to implement Ethernet services using the VPN technology over the IP/MPLS core (see Section 11) and MPLS-TP.

11 Virtual Private Technologies (VPNs)

VPNs refer to the combination of technologies to ensure communication between two points, through a "tunnel" that simulates a point-to-point connection inaccessible to the "eavesdropping" and interference. This private communication uses existing public networks like the Internet. Following the various types of VPNs are described.

11.1 GRE (Generic Route Encapsulation)

GRE [10] enables tunnels to be configured between source and destination routers as a point-to-point connection. Packets to be sent through the tunnel are encapsulated by a new header (the GRE header) that contains the end of the tunnel address. When the packets reach the end of the connection, the GRE header is stripped off and they continue to the destination determined by the original header. This type of VPN has disadvantages such as:

- GRE tunnels are usually configured manually, which requires a great effort in the management and maintenance.
- High processing required for packets encapsulation in GRE headers. A large number of tunnels may affect the performance of the network.

11.2 IP Security Protocol (IPSec)

IPSec allows authentication and integrity of each IP packet. It has two modes of operation: the transport mode, which only the payload of an IP packet is authenticated and encrypted, and the tunnel mode, which authenticates and encrypts the entire IP packet. Transport mode provides the protection of IP payloads such as TCP (Transport Control Protocol) segments, UDP (User Datagrama Protocol) messages or ICMP (Internet Control Message Protocol) messages. IPSec architecture is defined in [30].

11.3 MPLS VPNs

MPLS VPNs can be established by layer 2 or layer 3. In layer 2, the frames, such as Ethernet frames, are mapped on a virtual circuit MPLS (LSP), as shown in Fig.13.

The transport of layer 2 frames over MPLS networks is also known as Pseudowire Emulation (PWE) VPN, where different sites can share the same Ethernet network through the use of PWE VPNs. This mechanism is known as VPLS (Virtual Private LAN Service).

MPLS Layer 2 VPN is defined by the following IETF RFCs:

- RFC 4905: Encapsulation Methods for Transport of Layer 2 Frames over MPLS [35].
- RFC 4906: Transport of Layer 2 Frames over MPLS [36].

In MPLS Layer 3 VPNs, the IP/MPLS routers are virtually divided in several parts, so many as the number of users. The packets received at the user interface are inserted into the part exclusively allocated to each user. At the backbone, the traffic of each VPN is kept separated. The routing protocols such as OSPF/BGP-4 need to be associated with MPLS [49] to perform this VPN functionality, as shown in Fig.14.

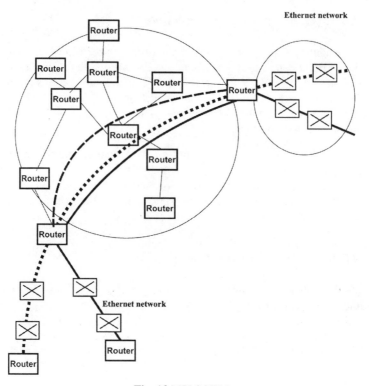

Fig. 13 MPLS VPN

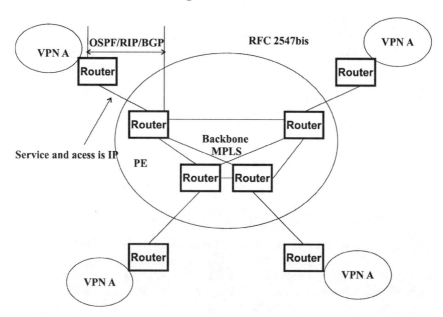

Fig. 14 MPLS Layer 3 VPN

Evolution of Internet Architecture 43

This type of VPN has advantages such as scalability, safety and facility of configuration when compared to MPLS VPN Layer 2.

12 Content Delivery Networks

Internet has been increasingly used as a platform to deliver content and a number of extensions are being proposed for efficient delivery and increased availability of digital contents to the Internet consumers.

Caching and replication are used to send digital content copies for key positions, close to the location of the user, to meet their requests quickly.

Caching creates temporary copies of part of the content such as Web pages, image files or multimedia files and intercepts user requests upon cache hits of the requested object. Then, the caching server sends the object to the user. Otherwise, the request continues its path to the source server. While caching is a technique to improve the transfer rate and to reduce the requirements for bandwidth, replication is a technique where the entire content is replicated to geographically dispersed servers. Replication (mirroring) techniques create and maintain content copies in a distributed way, under control of the providers. This is useful since the requests of the customers can be sent to the nearest server, reducing the dependence on central servers and serving the content more effectively.

CDN (Content Delivery Network) is a collection of servers arranged for more efficient distribution of digital content to end users. CDNs can be centralized, with a hierarchical structure or be completely decentralized. Typically, a CDN consists of an original server, proxy cache servers located at strategic locations, an infrastructure for handling user requests to calculate the best route in order to optimize the response time for the client and an infrastructure for billing. Fig.15 illustrates a typical CDN.

The CDN nodes cooperate to meet the user requirements, moving the digital content to others parts of the nodes distribution infrastructure ensuring high availability and good performance for the user. Compared to traditional caching techniques where ISPs (Internet Service Provider) use caching to pull digital content from a central server to the proxy servers nearest their customers, Digital Content Providers (DCPs) push the content to others distribution infrastructures overlaid to Internet. Then, these two approaches are complementary in delivering digital content as quickly as possible to end user.

The main research challenges in CDNs are in the multimedia distribution (video, audio, etc.) over the "best-effort" Internet infrastructure. During the 90s, research was focused on the development of multimedia servers to support the deployment of high-performance servers. Currently, researches have focused on cooperation techniques between CDN and Internet to solve infrastructure problems such as network congestion and overloaded servers. Studies have indicated that the number of continuous media stored in web-caching has increased significantly [44]. As a result, CDNs have become an intensive matter of research towards scalable architectures for multimedia caching.

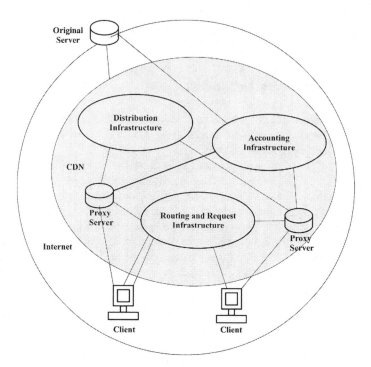

Fig. 15 Typical CDN

13 Peer-to-Peer Networks

P2P systems to file sharing are currently the most commons systems used for content distribution. In 2001, Napster was the application that grew fastest in the history of the Internet [44]. Since the advent of these systems, a large number of proposals such as Supernodes or decentralized approaches based on hash tables or tables scattering [44] have been developed. These tables have a special data structure, which combines search keys (hash) values. Using a simple key, it is possible to make a quick search and get the desired value. Hash tables are typically used for indexing large volumes of information (such as databases).

P2P networks enable the sharing of digital content among users ("peers") without a central control as the CDNs. These networks are characterized as information retrieval networks and are formed by ad-hoc aggregation of resources and by the collaboration among peers to locate, cache and/or or retrieve digital content and route the requests.

Peer-to-Peer networks are more failure tolerant and scalable than centralized systems, because they do not have a single point of failure. An entity can join or leave a P2P network anytime. These systems monitor files within the group of peers to promote reliable transfer of files and manage the heavy traffic caused by the high demand for popular files. These goals differ from the CDN goal in

Evolution of Internet Architecture 45

achieving the performance requirements of the network (e.g. response time), finding the nearest peer that has the desired content.

P2P management network systems are still very incipient, lacking for metadata, i.e. data containing information about data. However, these networks have leveraged the computing resources and disk bandwidth on the Internet and addressing a key issue related to digital content distribution: the large scale video distribution. In this case, the two main applications are: video streaming on demand (VoD) and delay sensitive multimedia applications such as Webseminars or broadcast TV over the Internet [44]. A large number of mechanisms have been proposed to solve the problem of multiple peers simultaneously transmitting and a receiving VoD. Related to multimedia live streaming, the solutions including the establishment of trees, similar to the IP multicast, tunneled in IP unicast connections. Protocols such as: SpreadIt, PeerCast, ESM, NICE and Zigzag [44] use this approach. Other protocols that treat multiple multicast trees are Splitstream, CoopNet and P2PCast. The extensive development of CDNs and P2P networks can change the current telecom operator's model from centralized control to a more decentralized control, where the techniques described above and new ones can be tested and adapted for telecom networks. There is a vast area of research and development to be explored in this sector.

14 IP Mobility and Multi-homing

Mobility and multi-homing are closely related: IP mobility and multi-homing protocols can be used together in a competing or complementary fashion.

IP mobility refers to a network functionality that maintains the same IP address to a node as the node move across different IP points of attachment. It allows users to preserve ongoing connections independently of the current IP point of attachment.

Multi-homing refers to a node (or a network) using multiple attachment points to the network to provide more reliability, redundancy, load balancing and so on. Arguably, "mobility is very fast multi-homing".

Typically in multi-homed network, a node has multiples addresses, what contribute to routing table explosion.

There are two types of multi-homing: host multi-homing and site multi-homing. In host multi-homing a host has more than one global IP address. These addresses may or not belong to different ISPs. In site multi-homing, a site has more than one Internet connection through one or several different ISPs.

Fig. 16 illustrates a site multi-homing example with two service providers (SP) and the different addresses divulgate to global Internet.

In case of SP1, the customer address is a subnet of SP1, facilitating the address aggregation but SP2, should divulgate to Internet the customer address and its own address (198.133/16) because they are different, contributing to explosion of the Internet routing table.

There are several proposals to address mobility and multi-homing to Internet and in the last developments, a Locator/Identifier separation concept that splits the current IP address into two separate spaces: one for host identifiers (Id) and other

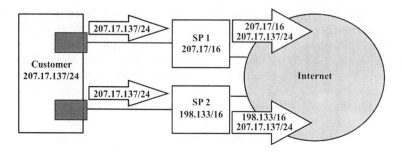

Fig. 16 Site Multi-Homing

for routing locators (Loc). This concept is similar to postal address where the person name is the identifier (never changes) and the locator is the residential address (can change).

Similarly, IPv6 auto-configuration characteristic, where the MAC address is inserted in the final part of the host address, when it is plugged into the network, could be used as a host identifier. Nevertheless, this mechanism failed as Loc/Id split concept because hosts treat the entire address as an identifier rather than the final part (MAC address).

There are basically two Loc/Id split solutions:

1. Loc/Id split handled by the host (host-based solution) where:
   a. IP address used only as a locator;
   b. Host identifier inserted in the IP packet as a header extension.
2. Loc/Id split handled by routers (router-based solution) where:
   a. The first/last routers add/remove a locator in the packets
   b. Require no hardware or software changes at the end-systems (hosts).

Some of the IP mobility/multi-homing protocols that use host-based approach are: LN6 (Location Independent Network Architecture for IPv6), HIP (Host Identity Protocol), SHIM6 (Site Multi-homing by IPv6 Intermediation), SIX/ONE and SIP (Session Initiation Protocol). LISP (Locator/ID Separation Protocol) and Six/One Router use the router-based approach.

In the following, we review remarkable approaches to handling mobility and multi-homing to Internet architecture.

14.1 MIPv4

In Mobile IPv4 (MIPv4) [43], when the mobile node is in its original domain, it is considered a conventional fixed terminal. However, when it traverses its original domain and access the Internet on a foreign network, it requests a "Care of Address" (CoA) of a Foreign Agent (FA) and registers this new address in its

Evolution of Internet Architecture

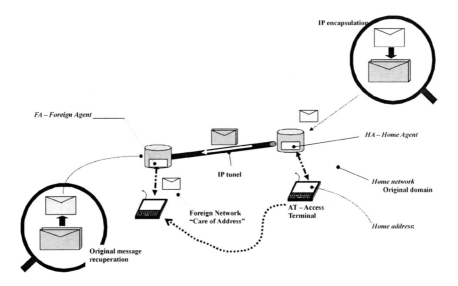

Fig. 17 MIPv4

Home Agent (HA) (see Fig.17). Thereafter, the HA intercepts all packets addressed to the mobile node, encapsulates them with a new header that contains the care-of-address forwarding them to FA. The FA forwards the packets to the mobile node based on the Home Address field.

14.2 MIPv6

Basically, MIPv6 [27], does not have the Foreign Agent (FA) functionality as in MIPv4. In MIPv6, when the mobile host visits a foreign network, it maintains its home address and receives a care-of-address valid in the foreign network. Then a binding is created between the home address and the care-of-address. The mobile node sends a message of binding update to its Home Agent (HA). HA, after receiving this message sends a binding acknowledgement to the mobile node, as shown in Fig.18a. Thereafter, the HA intercepts all packets from a correspondent node addressed to mobile node and forwards them to visited domain using a proxy functionality, as shown in Fig.18b. Then, mobile node replies directly to correspondent node, sending a binding update. The correspondent node starts sending the following packets directly to the address of the mobile node, as shown in Fig 18c.

Several problems related to Mobile IPv6 have been reported such as: the overhead of IPv6 extension headers, the location of the Home Agent is restricted to the home address of the node, the difficulty of replicating it on other network domains for fault tolerance, among others [52].

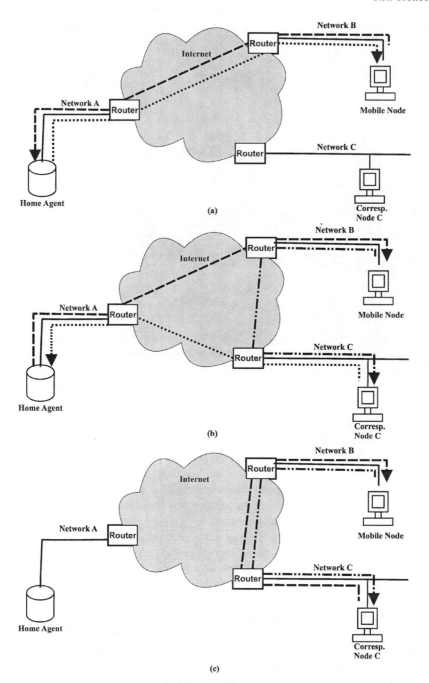

Fig. 18 MIPv6

14.3 SIP

SIP [9] is an application layer protocol developed by IETF group MMUSIC (Multiparty Multimedia Session Control) to provide advanced signaling and control for multimedia services over IP.

SIP establishes, modifies and terminates multimedia sessions, providing the means for addressing and location of its members. In session establishment, SIP acts as a signaling protocol with services similar to telephony signaling, but in the Internet context. Users can maintain the same identification, even if they plug in another attachment point of the network or use another device (personal mobility).

SIP uses URLs to provide the identification of its members and IP addresses are used as locator. The URLs can identify applications, users and flows. An URL example is tania@here.com;type=client;app=e-mail which identifies the user, the endpoint and the application.

SIP is a very mature technology and is largely deployed in the networks today.

14.4 LIN6

LIN6 [52] provides mobility support in IPv6 redesigning the IP address structure through the use of two types of addresses: the LIN6 generalized ID and the LIN6 address. The LIN6 generalized ID is as node identifier (ID), used as a name in the transport layer. It does not change even if the node moves. The LIN6 address contains both the node identifier (ID) and the node location (network prefix). It changes when the node moves to another network domain because the network prefix is different on each domain. Thus, the transport layers connections are provisioned using LIN6 generalized ID, while the network connections use LIN6 address. Nevertheless, to establish a connection at the transport level, it is necessary to know the current destination network address (and vice versa). Then, when the node moves to another network domain, it should register the network prefix in a Mapping Agent (MA) that makes the correspondence between the generalized LIN6 address (used at the transport layer) with the current LIN6 address (used by the network layer). Using this protocol, the node can move between networks domains maintaining its connections at the transport layer.

Advantages of this solution related to Mobile IPv6 are that it is possible to use multiple Mapping Agents, in a distributed way, and not only at Home Network and it do not require a special device to forward the packets. This protocol operates in mobile and multi-homing environments maintaining end-to-end connections.

14.5 HIP

HIP [38] provides a Loc/Id separation method that includes a Host Identity (HI) name space between the transport layer and the network layer. The TCP ports are connected to HIs and the HIs are dynamically connected to the IP addresses, enabling network address changes during established sessions. Wherever a host needs to change its address, it sends an *update* message, to inform the new location to its peer. HIs are public keys and are negotiated prior of the establishment of the communication. That way, HIP provides secure communication channels between the peers. In [39], the specifications of this protocol are detailed.

14.6 New Proposals

14.6.1 SHIM6

Shim6 [42] is a new multi-homing proposal that aims to eliminate the contribution of multi-homing multiples addresses to the global routing table explosion. It provides a site-multi-homing solution for IPv6 preserving the global routing scalability. The multiple globally routable prefixes are obtained from each correspondent ISP, avoiding that the site addresses to be divulgated and added into the global routing table. Fig.16 illustrates the global routing scalability problem.

Shim6 introduces an *Upper Layer Identifier* (ULID) that persists at transport layers and above, while different addresses are exchanged at network layer based on the mobile node location. A shim sub-layer is located within the IP layer, between the IP endpoint sub-layer and the IP forwarding sub-layer, as shown in Fig. 19.

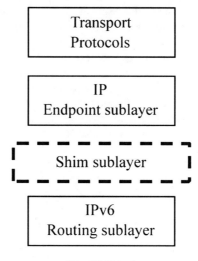

Fig. 19 Shim6

Shim6 protocol manages the binding *ULID/locator* for the peers and provides failure detection and route protection mechanisms.

14.7 SIX/ONE

SIX/ONE [55] is a host-based Loc/Id split solution for IPv6 where each host has multiples IPv6 addresses from different providers (multi-homing) being that these

Evolution of Internet Architecture

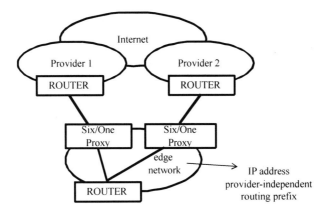

Fig. 20 Six/One

addresses share the same last 64 bits (MAC address). SIX/ONE proxy at edge network hides these addresses and translates them for a provider-independent routing prefix, as shown in Fig.20.

SIX/ONE is a scalable and backwards compatible solution for multi-homing [55]. However, to eliminate renumbering in the edge networks, Six/One router-based approach were designed.

14.8 LISP

LISP [7] is a new ongoing in IETF standard Loc/Id router-based solution that uses a tunneling technique to Loc/Id split. The main idea is to insert a new IP layer under the current IP layer, as shown in Fig.21.

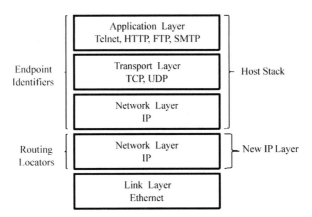

Fig. 21 New IP Layer

This layer contains addresses referred as Routing Locators (*RLOCS*) used for routing the packets through the IP network. The original IP address is maintained unchanged at the endpoints and is used as an endpoint identifier (*EID*) and to local routing. For inter-domain routing, the *EIDs* are mapped on *RLOCs* by the edge router. Then, the packets are forwarded to the destination edge router where the *RLOCs* are stripped off the packet and it is routed based on its *EID*.

The *EID/RLOC* mapping methods are under study. To implement this solution, it is necessary a lot of modifications on the routers, being necessary a gradual implementation on the Internet.

15 Discussion

Extensions to the Internet have been developed at increasing rate. Above, we discussed CIDR, IntServ, DiffServ, DHCP, DNS, NAT, MPLS, VPNs, IPSec, CDNs, Mobile IP and Multi-homing. However, many of these extensions were developed as individual solutions for specific problems without an architectural approach. Next, some inconsistencies between the extensions and the principles of the Internet architecture will be discussed.

As Internet growth with introduction of Web, two important techniques were developed to reduce the rate of addressing consumption: NAT and DHCP. These two approaches violate the Internet principles of use a unique and globally valid address for each host and the transparency (end-to-end argument). NAT uses private addresses not visible globally and the address translation breaks end-to-end connectivity. Dynamic assignment of addresses by DHCP, assigns different addresses to a given host each time it is connected to the network. The addresses are usually stable for a time period but can be reassigned for the ISP. Then, NAT and DHCP prevent use hosts in environment in which other hosts (peers) needed to contact them directly such as peer-to-peer networks. They built a barrier to move Internet towards a decentralized peer-to-peer model. Some mechanisms have been proposed to traverse NAT, namely NAT transversal (NAT-T) and especially used in peer-to-peer, VoIP and online games. Some NAT-T protocols are STUN and TURN [37]. STUN is not a self-contained NAT traversal solution applicable in all NAT deployment scenarios and does not work correctly with all of them. TURN allows the host to control the operation of the relay and to exchange packets with its peers using relay. This protocol has been standardized in RFC 5766[32]. SIP also does not work through NAT unless NAT are extended with NAT-T.

Nevertheless, NAT and DHCP are very useful for facilitate renumbering and NAT provides security. IPSec is fully consistent with the simplicity principle of Internet, encrypting the information at the transport layer (end-to-end argument) but do not function properly with NAT, neither with caching mechanisms.

Firewalls also violate the end-to-end argument because hosts behind a NAT/Firewall can not receive incoming traffic initiated by a foreign host.

On the other hand, DNS provides hosts identification via domains and was the first step towards an identification/localization decoupling necessary for IP mobility.

Evolution of Internet Architecture

IPv6 aims to restore the Internet architectural principles of simplicity and transparency, by increasing the addressing space and eliminating NAT. Moreover, it adds auto-configuration functionality to IP, making it "plug-in-play" and together IPSec provides security. Nevertheless, to adopt IPv6 is painful, expensive and requires mechanisms for IPv4 and IPv6 interoperability and other protocols v6.

IntServ and DiffServ violate the Internet principle of simplicity, making IP routers job harder, increasing the IP layer processing and complexity.

MPLS is an ambiguous solution because it adds a connection oriented functionality to IP layer and, associated with Ethernet, mostly reinvent what IP does, at layer 2, becoming a competitive solution. Moreover, it violates the Internet architectural principle of connectionless architecture (best-effort connections) and changes the layered reference model (MPLS works at 2.5 layer).

Finally, the IP mobility/multi-homing solutions solve the overload semantic of IP address making an identifier/localization split enabling host/node mobility and avoiding routing table explosion. The caveat of the solutions is requiring changes in the original layered reference model introducing new layers between transport and network layer. Moreover, LISP introduces a "new Internet layer". The scalability and security aspects of these solutions need be estimated.

In summary, nowadays Internet suffers with loss of transparency, scalability problems, incompatibility among protocols, protocols taking roles for which they weren't originally designed, security vulnerability and attacks, new layers have been added between the existing ones (MPLS (L2.5), HIP (L3.5), SHIM6 (L3.5) and LISP (L3). Fig.22 shows the current form of the Internet waist.

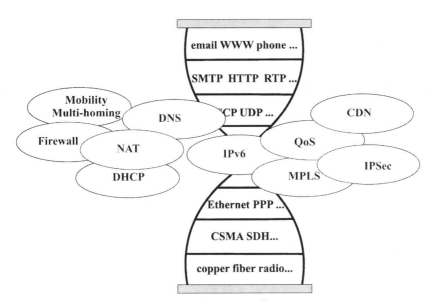

Fig. 22 Current Internet Waist

It is now time to rethink the Internet architecture and re-engineering it to address the current and future requirements. Next-generation Internet architectures research projects have been appeared everywhere to dealing with this situation of restore the architecture coherence. The first steps for this re-engineering process are to design evolution scenarios and to define the new requirements. These steps will be discussed in the next chapters.

References

1. Anderson, L.: RFC 5317 Joint Working Team (JWT) Report on MPLS Architectural Considerations for a Transport Profile (2009)
2. Blake, S., et al.: RFC 2475 - An Architecture for Differentiated Services (1998)
3. Braden, R., Clark, D.: RFC 1633 - Integrated Services in the Internet Architecture: an Overview (1994)
4. Braden, R., et al.: RFC 2205 - Resource ReSerVation Protocol (RSVP) Version 1 Functional Specification (1997)
5. Deering, S.: RFC 2460 - Internet Protocol, Version 6 (IPv6) Specification (1998)
6. Droms, R.: RFC 2131 - Dynamic Host Configuration Protocol (1997)
7. Farinacci, D., et al.: Draft-ietf-lisp-06.txt- Locator/ID Separation Protocol (LISP) (2010)
8. Fuller, V., et al.: RFC1519 - Classless Inter-Domain Routing (CIDR): an Address Assignment and Aggregation Strategy (1993)
9. Handley, M., et al.: RFC 2543 - SIP: Session Initiation Protocol (1999)
10. Hanks, S., et al.: RFC 1701 Generic Routing Encapsulation, GRE (1994)
11. Hinden, R., Deering, S.: RFC 2373 – IP Version 6 Addressing Architecture (1998)
12. IEEE802.1ad, Provider Bridges (2005)
13. IEEE802.1ah, Provider Backbone Bridges (2008)
14. IEEE 802.1D, MAC bridges (2004)
15. IEEE802.1Q, Virtual LANs (2006)
16. IEEE802.1Qay, IEEE standard for Provider Backbone Bridges Traffic Engineering (2007)
17. IEEE802.1w, Rapid Reconfiguration of Spanning Tree (2004)
18. IEEE802.1s, Multiple Spanning Trees (2006)
19. IEEE P802.3ba, 40Gb/s and 100Gb/s Ethernet Task Force (2007)
20. ITU-T Recommendation G.872, Architecture of optical transport networks (2001)
21. ITU-T Recommendation G. 783, Characteristics of synchronous digital hierarchy (SDH) equipment functional blocks (2006)
22. ITU-T Recommendation.G.8110, Architecture of Transport MPLS (T-MPLS) layer network (2006)
23. ITU-T Recommendation. G.8112, Interfaces for the Transport MPLS (T-MPLS) hierarchy (2006)
24. ITU-T Recommendation G.8121, Characteristics of Transport MPLS equipment functional blocks (2006)
25. ITU-T Recommendation Y.1731, OAM functions and mechanisms for Ethernet based networks (2006)
26. Jiang, S., et al.: draft-ietf-v6ops-incremental-cgn-00.txt - An Incremental Carrier-Grade NAT (CGN) for IPv6 Transition (2009)
27. Johnson, D., et al.: RFC 3775 Mobility Support in IPv6 (2004)
28. Katsube, Y., et al.: RFC 2098, Toshiba's Router Architecture Extensions for ATM: Overview (1997)

29. Katz, D., Ward, D.: draft-ietf-bfd-multihop-09.txt - BFD for Multi-hop Paths (2010)
30. Kent, S.: RFC 4301 Security Architecture for the Internet Protocol (2005)
31. Koren, D.: ATM Tutorial (2007),
    http://www3.rad.com/networks/infrastructure/atm/main.htm
32. Mahy, R., et al.: RFC 5766 - Traversal Using Relays around NAT, TURN (2010)
33. Mankin, A.: RFC 1752 - The Recommendation for the IP Next Generation Protocol (1995)
34. Mannie, E.: RFC 3945 Generalized Multi-Protocol Label Switching (GMPLS) Architecture (2004)
35. Martini, L., et al.: RFC 4905 - Encapsulation Methods for Transport of Layer 2 Frames over MPLS Networks (2007)
36. Martini, L., et al.: RFC 4906 - Transport of Layer 2 Frames Over MPLS (2007)
37. Matteus, P., Rosenberg, J.: RFC 5766 - Traversal Using Relays around NAT (TURN): Relay Extensions to Session Traversal Utilities for NAT, STUN (2010)
38. Moskowitz, R., Nikander, P.: RFC 4423 - Host Identity Protocol (HIP) Architecture (2006)
39. Moskowitz, R., et al.: RFC 5201 Host Identity Protocol (2008)
40. Newman, P., et al.: RFC1953 - Ipsilon Flow Management Protocol Specification for IP (1996)
41. Nichols, K.: RFC 2474 - Definition of the Differentiated Services Field (DS Field) in the IPv4 and IPv6 Headers (1998)
42. Nordmark, E., Bagnulo: RFC 5533 - Shim6: Level 3 Multi-homing Shim Protocol for IPv6 (2009)
43. Perkins, C.: RFC 3344 - IP Mobility Support for IPv4 (2002)
44. Plagemann, T., et al.: From Content Distribution Networks to Content Networks – Issues and Challenges. Computer Communications 29(5) (2006)
45. Rekhter, Y., Li, T.: RFC 1518 - An Architecture for IP Address Allocation with CIDR (1993)
46. Rekther, Y., et al.: RFC1918 - Address Allocation for Private Internets (1996)
47. Rekhter, Y., et al.: RFC 2105, - Cisco Systems Tag Switching Architecture Overview (1997)
48. Rosen, E., et al.: RFC 3031 - Multiprotocol Label Switching Architecture (2001)
49. Rosen, E., et al.: RFC 2547bis - BGP/MPLS IP VPNs (2004)
50. Shah, S., Yip, M.: RFC 3619 - Ethernet Automatic Protection Switching, EAPS (2003)
51. Srisuresh, P., Egevang, K.: RFC3022 - Traditional IP Network Address Translator (2001)
52. Teraoka, F., et al.: Internet Draft (draft-teraoka-mobility-lin6-00.txt) - LIN6: Mobility Support in IPv6 based on End-to-End Communication Model (2000)
53. Tsirtsis, G., Srisuresh, P.: RFC 2766 - Network Address Translation - Protocol Translation, NAT-PT (2000)
54. Viswanathan, A., et al.: Internet-Draft - ARIS: Aggregate Route-Based IP Switching (1997)
55. Vogt, C.: Draft-vogt-rrg-six-one-02 - Six/One: A Solution for Routing and Addressing in IPv6 (2009)

Scenarios of Evolution for a Future Internet Architecture

Tania Regina Tronco[1], Takashi Tome[1], Christian E. Rothenberg[2], Marco A. Ongarelli[1], and A.C. Bordeaux Rego[1]

[1] CPqD Foundation, Rodovia Campinas Mogi-Mirim, km 118,5,
Campinas – São Paulo, CEP 13096-902, Brazil
{tania,takashi,onga,Bordeaux}@cpqd.com.br
[2] University of Campinas (UNICAMP), Cidade Universitária "Zeferino Vaz"
Distrito de Barão Geraldo - Campinas - São Paulo, CEP 13083-852, Brazil
chesteve@dca.fee.unicamp.br

Abstract. This chapter describes three scenarios for a future Internet architecture: the user-centric, the object-centric and the content-centric. These scenarios are neither "different" nor "mutually exclusive" and they contain some predictions about the network usage evolution in our diary life and to do business. As an important part of the research on new architectures for the Future Internet, these scenarios are characterized by their network's attributes and exemplified with some use cases. The identification of the attributes is one step toward the new Internet architecture requirements. During the research, some key convergence aspects among the scenarios were identified as well some specific characteristics of each one.

1 Introduction

The technical requirements for the Internet have changed considerably since end of seventies when it was conceived. At that time, there were about 200 hosts connected to ARPANET. Nowadays, this number increased significantly to more than 500 million [15] and it is expected to increase more and more. This connectivity explosion occurred mainly with the increase development of new Web applications, after Internet became commercially exploited, in 1991. Currently, there is a consensus that the Internet evolution will cause a strong impact on the economy and way of life in society consisting of a powerful driver for innovation, economic growth and social development. These aspects have been the focus of many global discussion forums, such as the Internet Governance Forum (IGF), International Institute for Sustainable Development (IISD) [5] and others. The pervasive computing, through sensors and mobile devices will be part of people's daily life enabling relevant electronic services such as e-health, remote medical care, environmental monitoring, e-government, social networking, improvement our quality of life in general. Moreover, universal access to information and communication, mobile devices and services available "anywhere/anytime" might reduce the digital divide. In this context, the network will be essential for people and in businesses, creating a lifestyle networked-centric with the increasing online presence of companies and new service providers. The Service Oriented Architecture (SOA) and its instantiation through the Web services technology has enabled the development of new applications in diverse areas such as financial, commerce, content distribution, network management, etc. In keeping this trend, mobility, portability and ubiquitous access to data and services become more important than the terminal access itself. This is an example of how technologies can change paradigms and, after the powerful mainframe computers era, came the networks era.

In this context, the education will also be influenced by the right of access to the network and a new social division can be created based on the type of Internet access that each citizen will have [5]. Government policies must be developed to provide legal and regulatory issues such as network neutrality, privacy and digital rights.

Therefore, treating the future of Internet and to draw evolution scenarios including different possibilities of rupture are an important steps for thinking not only about the future, but especially to take decisions now to achieve the most desired future scenario.

From a technical viewpoint, there is a consensus among researchers and engineers involved with network technologies that the Internet architecture should evolve to support the expansion of new services and the traffic demand in the coming years. The main issues involve aspects such as mobility and ubiquity, security and capacity. The importance of the so-sought new architecture for the Internet is emphasized by the large number of projects worldwide designing and testing new solutions to achieve the new expectations of network evolution.

Nevertheless, there are diverse approaches on how to promote this network upgrade. In ARCMIP project[1], we consider a top-down approach (see Fig.1) that consists in: (a) describe a set of evolution scenarios, containing predictions of various modes of network usage in our daily life and to do business, (b) map these scenarios in three categories (A, B and C) with its attributes and challenges to the current Internet architecture, (c) use the attributes to derive requirements for the new network architecture and (d) employ these requirements to build recommendations and to identify important research themes to be developed next years.

The attributes identified in (b) are an important part of the research because they may reflect some convergence aspects among the scenarios and can help to identify new requirements for the Future Internet evolution. Nevertheless, the requirements identified in the top-down approach are not exhaustive to design a new Internet architecture because they can have a tendentious viewpoint with a regional influence and, at same time, they must be applied globally. Then, the new Internet architecture must be open and flexible to deal with lists of requirements coming from different regions of the world and with different level of importance

---

[1] ARCMIP – ARchitectures for Mobile IP – is a project to explore new technologies for Future Internet sponsored by FUNTTEL (Funding for Technological Development of the Telecommunications) - Ministry of Communications, Brazil and currently in development at CPqD Foundation.

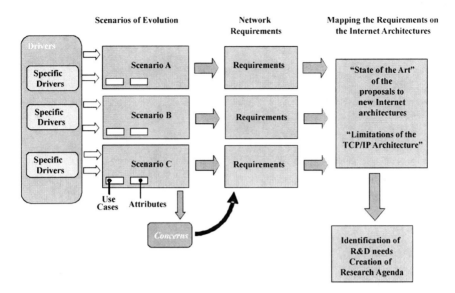

Fig. 1 Future Internet Research Top-Down Approach

for each one [2]. Hence, Internet probably will not be unique, but multiple federated Internets build with different design criteria will coexist.

2 Scenarios of Evolution

There are several possible angles from which the future Internet can be looked at. In our context, "main scenarios" are those which have a huge impact (i) on traffic volume and patterns, (ii) on the physical and virtual network topology and (iii) on daily-life of citizen that becomes the centre of attention of the future Internet projects. In other words, we are interested in those scenarios that are beyond a linear bandwidth-growth, or ad-hoc solutions, for more important they can be.

After analyzing several candidate scenarios, we concluded that three of them could have a huge impact on the network architecture and aggregate functionally most of the envisioned use cases: (a) a user-centric scenario, (b) a content-centric scenario and (c) an object-centric scenario. There is a lot of work being done worldwide on those scenarios separately; however, they are neither "different" nor "mutually exclusive" futures. Rather, we consider they are part of a holistic, more complex future, from which we can see only some selected views, as shown in Fig. 2.

The user-centric scenario is related to provide a ubiquitous and comfortable service portfolio (communication, seamless mobility, ergonomics, trust, no wait times, etc.) to the people themselves. The content-centric scenario refers to convert the network nature itself, from a "link-structure" (set of a sparsely distributed nodes connected through links) to a "network of information", i.e., a receiver-driven information dissemination infrastructure, with information objects

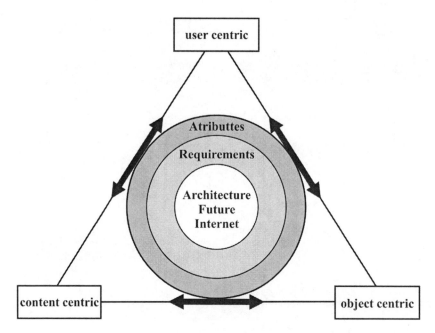

Fig. 2 Future Internet Evolution Scenarios

uncoupled from specific hosts locators. Its associated paradigm would be an "Earth-wide thinking machine" (intelligence spread everywhere like inside the human's brain).

The object-centric scenario opens the Internet-scale connectivity to any imaginable real world object, expanding the host and device endpoint space to sensors and things. The so-called Internet of Things is based on the paradigm of "everything is a living and intelligent being", from clothes to streets.

The first scenario (user-centric) was chosen because, whatever else we do, human communication continues being the centre of our attention. The second scenario (content-centric) appears to have the hugest impact in changing the very nature of what internetworking is actually about and the third one (object-centric) is a "brave new world" with the potential of changing the physical world itself, where we live in and to have a hugest impact on the scalability of the network.

Next sections, these scenarios will be detailed in terms of attributes that best describe them and some use cases are presented, for the sake of illustration of the dominating features of the scenarios. Finally, a brief discussion about the common and specific aspects of each scenario is presented in section 4.

2.1 User-Centric Scenario

The user-centric scenario goal is to provide a ubiquitous, comfortable and personal services portfolio to the people. It is characterized by high user interactivity with the network, through a wide variety of interfaces and by the personalization of the

Scenarios of Evolution for a Future Internet Architecture   61

services to the user's preferences and their contexts. The expectations of the future Internet, from the user-centric viewpoint, are that the network interactivity is quite simple and secure, i.e., users want only enjoy the network, in a simple and confidential way. The main characteristics of this scenario is that the new services can be composed and harmonized according to the users' instantaneous needs to enrich their daily life and to meet diverse lifestyles.

This scenario encompasses a wide variety of network technologies and devices that can be dynamically selected by the users, depending on their current needs. The device interfaces must provide the appropriate ways for a rich user interaction, i.e., users should be able to configure their own services through the Web access and catalogs available on the Internet. The human behavior may also may affect the use of network resources [7] as when a user always visits certain areas more than others, these motion pattern can be used to predict the transport of messages to a specific receiver or a group of people with similar interests can form a community and be found more frequently if a community-based routing could be built. Furthermore, users should be able to produce/share/export digital contents and services becoming at same time producers and consumers of digital content and services. The interaction between users and the digital content has increased. The Web has reemerged as a powerful medium for personal expression and community, through distribution of user-generated digital content (UGC). Currently, UGC has been composed by blogs, community sites like Slashdot, Flickr, Wikipedia, YouTube, and MySpace friend's network, among others, where users insert videos, photos and other audio-visual contents. The goal of Web 2.0 is becoming a conceptual infrastructure and technology in which the UGC can proliferate. The introduction of semantic Web (Web 3.0) will contribute significantly to the interpretation and analysis of user preferences. Metadata, i.e., data about data, should be aggregate to digital content to provide context and purpose, allowing describing, organize, systematize, manage and edit it. Then, metadata can be used for various user-centric functions including:

- Describe the user's preferences and history of use;
- Describe the conditions for access to content: authentication, intellectual property, copyright and costs;
- Content Advisor in predetermined categories;
- Define the context in which the content was created.

The context management via ubiquitous computing techniques such as "context-aware" can adapts its operations to the user's moment context without explicit user intervention and takes into account the environmental conditions. The context information can be acquired from a variety of devices such as: sensors, network information, browsers, etc. Moreover, recent advances in 3D processing will enrich user's immersive media experiences and create a virtual reality with a world of virtual digital objects. Sensors will take part of this new world to get information about smell, taste, emotions, etc. [20].

In user-centric scenario, people can make use of RFID (Radio-Frequency Identification) chips implanted and activated at different levels of security to facilitate medical aid, to make medical procedures low costs (online life), to security, etc. In this context, the quality of connection (in terms of speed, reliability) and the practice of multi-homing, which the user is connected to more than one service provider, is an important requirement.

To wrap up, this scenario requires the organization of services and applications around the user's life with minimal human intervention (autonomous systems) and requires continuous monitoring of the user's environment, its context and the available network resources.

The main attributes of this scenario are:

- User's mobility and ubiquity;
- Dynamic services user's controlled;
- Context-awareness and content-aware networks;
- Dynamic auto-configuration of the network;
- Users acting as consumers and creators or distributors of 3D digital content;
- Social communication and interaction;
- Multi-homing;
- Security and privacy;
- Personalized advertisements and
- Search tools and identity management of virtual digital objects.

2.2 Object-Centric Scenario

This scenario refers to a worldwide network of interconnected objects, each one with a unique address, communicating via Internet protocols suite.

This scenario involves a wide variety of objects totally different in terms of functionality, technology and scope, with labels RFID and embedded sensors, connected to the Internet. The current generation of Wireless Sensor Networks (WSNs) consists of a collection and aggregation of measurements data from of a large number of sensors in a specific geographic area. In the near future, sensors network will include systems, sensors/actuators forming a closed circuit for real time control of physical objects. Essentially, the idea of the Internet of Things is that all (or almost) objects have intelligence (chip) and an identity (IP) itself and almost of the interactions between objects that are currently enabled or mediated by human beings will be automatically made by the objects themselves. Additionally, this scenario considers a few things that today are futuristic (or at least there are not in large-scale) like robots of various sizes running various functions, since cars to automatic domestic vacuum cleaners or even smaller things, characterizing the so-called wearable computing.

The possibility of direct communication between objects without human intermediation has a large technical, economic and social potential [9; 19] and is named as "Internet of Things".

Although the basic concept of "Internet of Things" is the same since its conception in 1998, the perception of its meaning - and therefore the extension of the concept - has changing over time, as shown in Figure 3.

Scenarios of Evolution for a Future Internet Architecture

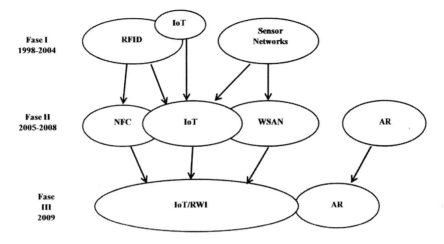

Fig. 3 Conceptual Evolution of the Internet of Things

The original idea of Internet of Things was closely linked to RFID concept [18]. Originally, the RFID tags were used to enable robots to recognize objects. In the meanwhile, RFID tags evolved to act not only as passive labels as a bar code, but as devices containing some intelligence, leading to the concept was changed to Internet of Objects, in which objects could communicate. As an example, a smart refrigerator could know how many beer cans were inside it. Therefore, Phase I of the Internet of Objects is characterized by this vague notion of communicating objects, without a specific goal.

The next phase was the "sensor network" [8; 12; 1], characterized as a combination of a large number of sensors, generally sending small amounts of information to a central processor. A sensor networks is a network with a large amount of nodes in motion, transmitting information with a low error tolerance and in real time. Additionally, each node is usually a small device with low processing power, low energy availability and low power RF transmission.

In Phase I, which runs from 1998 to 2004, the concepts of RFID and sensor network were not linked. In 2005, these concepts were unified under the notion of an Internet of Things (IoT), a set of animated objects communicating with each other [9]. Another conceptual expansion occurred during this period was that the short distance wireless communication left the exclusive RFID domain and started to embrace new opportunities such as to be used in cards which would communicate through techniques of NFC (Near Field Communications), as, for example, in the mobile phones. Therefore, Phase II is characterized as a world composed of intelligent and communicating objects.

Although the technological challenges and potential IoT applications are enormous, in 2009, it climbed a new conceptual level, called Phase III. This new phase is characterized by incorporating two new concepts. The first, called the Real World Internet (RWI) [17] that consolidated the sensors/actuators network as its basis. The IoT is a set of communicant's objects via Internet and through Internet and sensor devices, a person can get information about anything, from anywhere on world in real-time. In this sense, the Internet becomes the interface through which a person can interact with the people and objects in any part of the world. The second concept embedded in IoT is the "Augmented Reality" (AR). AR is originated from virtual reality technology and mix in real time, real-world information with the virtual-world information. For example, an AR image can include both, a people photograph and an object image mixed in real time. This new concepts makes the Internet more than a communicant interface, but a complex universe in which the real world and virtual world are connected and confused.

The main attributes of this scenario are:

- Identification of a large number of mobile and a ubiquitous objects;
- Interaction among objects without human interference (smart objects);
- Growth of robots and machines in the domestic environment;
- Increase of real time traffic to control the objects and device and
- Security, privacy and reliability of control information.

2.3 Content-Centric Scenario

The current Internet was based on a host-centered communication and this model has changed toward a network for distributing and sharing digital content. Currently, the majority uses the Internet (more than 95%, according to Van Jacobson [10], is to get a "piece of data identifiable with a name" (named chunk of data). This piece of data can be the content of a Web page (image, video), a multimedia content (movies, music), an instant message or a package with a fixed content. The main goal of communication between machines is to acquire contents with a specific name, e.g. an URL. Digital content stored and delivered via the network (online) became a new medium that has the potential to reach a quality level different from other current media (TV, radio, etc.) because:

- It can be delivered instantly to millions of people geographically dispersed;
- It can be stored and displayed ("displayed") in various types of access devices such as computer monitors, cell phones, PDAs, interactive TV, etc.;
- It enables interactivity and can be changed dynamically, adapting to user behavior.

With the emergence of this new type of media, new demands, opportunities and requirements appear on several levels, such as production of new digital content, distribution (delivery), access and new consumer markets.

In this context, the idea of content is completely detached from any notion of location (address), which demands new functional requirements for routing and storage from the network infrastructure. The user is concerned only with the authenticity and timeliness of the digital content, being indifferent to their location.

The Internet limitations to address access and massive distribution of content, beyond the conflicts of interest in the business models (tussle networking), have motivated the emergence of the new proposals for service-oriented networks [6] data networks [13] and content networks [11] and they are rapidly gaining importance [16].

These new paradigms are searching for a network to enable:

- Efficient access and distribution of information objects (e.g. multicast techniques, in-network caches);
- Avoid heavy and disruption intolerant operations such as end-to-end connections establishments;
- Allow the data transport in an opportunistic way (e.g. Ad Hoc and Delay Tolerant Networks).

In this context, proposals for content-based routing have been developed. Content-based routing is based on receivers expressing their interest in specific content by sending messages through the network or by subscription [4]. The network is responsible for delivering to each receiver the desired contents. A content-based network [3] is an infrastructure of communication in which the flow of messages between senders and receivers is determined by characteristics expressed (and controlled) by the receiver, rather than the knowledge of the destination addresses.

Recent developments in Information Technology (IT) also confirm the shift in focus from a communication networks between hosts to information-centric network. The traffic engineering of these networks has also changed from flows of packets to services, including the notion of semantic Web. This trend on services is accompanied by an emergence of applications and service-oriented architectures (SOA) and infrastructures (SOI).

In summary, most Internet applications are targeted to data (content) and services, regardless of their location by means of DNS redirection and anycast type of services. The native support of the "information objects" (a general term used to data/content/service) in the network is desirable to increase the efficiency and possibilities of content management. This may lead to new Internet operation paradigms such as "publish-subscriber" where the main idea is that content is completely disconnected from its location [14]. In this paradigm, the users express their interest in pieces of content without worrying where it comes from, since they are able to authenticate it and have guarantees that it really corresponds to the desired content.

In addition, cloud computing continues to consolidate around the Internet of the Future as a global platform for applications and storage of user's data. The research in this area may foster the development of new protocols and (data center) network architectures that unify the network access and data storage in a single and efficient power consumption way. Objectives include CAPEX/OPEX reduction by simplifying the number of network adapters, switches, cables, power requirements and cooling and also save resources of man-hours needed for setup, maintenance and expansion of infrastructure. Moreover, the interconnection of geographically distributed data centers may lead to new overlays (Inter-Cloud) on the public Internet such as evolved CDNs (Content Delivery Networks).

This information-oriented and distributed services scenario can influence the Future Internet architecture in a decisive way. Services with an emphasis on content will increasingly be incorporated to serve traditional users connected to the network via wired access, full mobile users (user, terminal, session, domain), collaboration (ad-hoc networks), sensing (sensor networks) and usability, the latter with emphasis on aspects related to the ergonomics of easy access to content and information objects.

The main paradigm of this scenario is "sharing my information anywhere, anytime" and its main offenders are the problems related to privacy, scalability and transition of the current network infrastructure to network architecture able to provide resources and tools to support the distribution of all these content in appropriate way.

The main attributes of this scenario are:

- Creation and efficient dissemination of digital content;
- Identity of information items;
- New content routing paradigms;
- Receiver-driven networking (e.g. publish/subscribe);
- In-network caching and replication;
- Network enablers for localization and search of information objects;
- Multicast, anycast and multipath information flows and
- Security and privacy.

3 Use Cases

Next, three use cases are presented, for the sake of illustration of the dominating features of each scenario: user-centric, object-centric and content-centric. These use cases are just representative and their presentation has the main objective of highlight some possibilities of Internet usage in the future and illustrate the attributes presented in each scenario.

3.1 Use Case – User-Centric Scenario

The use case presented here refers to an executive´s life of the future improved by the Internet usage.

Paul, an executive, often is in business travels and exchanges information with other executives and his family via 3D Media Internet. He has a virtual identity card and a correspondent avatar. New interfaces allow the capture and the exchange of taste, smell, emotions' information and enrich his communication.

Paul often forgets locations and schedules of bus stops, travel routes, flight schedules and is reminded of his agenda for his avatar. The personalization of the services through the aggregation of multiple sources of information such as calendar, flight plans, geographic location enable him establish a personal agenda and being automatically notified about his commitments. Paul is also frequently advertized about offers in his shopping list and, during some waiting times at the airport, he plays online games and meets people in social networks. While Paul is at the airport, receives an emergency phone message from health monitoring system of his mother. He calls her immediately and discovers that she's okay. Cellular phones are integrated with health monitoring systems in real time by Internet, providing health care services with personalized alerts, medications, diets, etc.

3.2 Use Case - Object-Centric Scenario

The use case presented here refers to a housewife's supermarket shopping. It presents nothing "breaking new" from previous works but tries to bundle the several proposals, analyzing what makes sense (or not).

The use case starts when the housewife/houseman is in the kitchen, after breakfast, preparing the daily agenda before going to supermarket. As usual, (s)he may have the TV set on, which can be presenting some recipe.

If she got interested on the recipe, she can, instead of annotating it by hands, simply send a "copy" command through her PDA (acting now as TV remote control) to the TV set, to have the recipe copied onto PDA. In any digital TV system, a set of video, audio and data information are bundled together (i.e. multiplexed) to form a program stream. At the receiver side, the information flows are demultiplexed and presented. Some information may be kept in a hidden mode (or even not processed), until the user requests it, as happens with SAP audio or textual closed caption. In our use case, the recipe has a hidden text version, together with the presenter's audio/video, which is kept hidden for a while. When the user sends the "copy" command, either the entire multimedia "how-to-do", or only the textual part of the recipe, is copied onto the PDA, either directly (e.g. via short-range connectivity like Bluetooth) or by means of a "domus server", which is an extension of the home gateway and acts as the rendezvous point for the "family information network".

In the next step, the domus server will check the availability of the recipe's ingredients, communicating with kitchen boxes, containers, and so on, to find the requested RFID tags. Then, the missing items will be added to the "shopping list". The shopping list itself is an information object from a content-centric perspective, with metadata information linking the unique identifiers of the information objects in the shopping list. It is once prepared collectively by the family members (maybe including some machines as "family members") and, after that, updated automatically as needed. Both PDA and the domus server have a copy of (or access to) the shopping-list information object.

When the housewife is ready to go shopping, she pushes the command "verify-supermarkets" in her PDA. Then, the PDA sends the shopping list to near-by supermarkets. After some seconds, the supermarkets' support center will return the list checked, pointing out the prices, missing items and today promos. After analyzing the promos, the user is able to send a message to the selected supermarket informing about her/his willingness to purchase the selected items. Then, the supermarket server (operation center) will reserve the selected items, guaranteeing them and avoiding the customer dissatisfaction of not finding the desired items when (s)he arrives. To avoid "false reservoir", the supermarket can do such reservation based on the customer registry, or a trusted third party identity manager (e.g. a telecom operator).

The housewife/man uses her/his eco-car (another use case) to drive to the selected supermarket. Soon the way, the PDA's shopping list can be remotely updated by the domus server, because the washing machine warned late that soap run out.

Shopping in the future supermarket can be slightly different from today's one. It can look more like an indoor garden than like today's warehouse appearance. Inside it, boulevards will have plants, small tables and chairs, and, of course, boots of cans and boxes. The user can shop pushing a physical cart as today, getting and putting things into the cart, or simply taking an electronic cart at the entrance and annotating the desired items (reading the item RFID with the cart) as (s)he walks by the boulevards and sees interesting items. Alternatively, this can be done with the user's own PDA, assumed the required supermarket's software is installed or accessible via online services. In this case, when the user arrives at the supermarket, (s)he simply registers a "new shop" and gets a onetime shopping ID.

Both the physical cart, equipped with a display, and the tablet-size electronic cart, can help the shopper not only with the shopping list, but also exhibiting additional information like nutrition facts, "how-to-use" hints, recipes, or even bringing to the item manufacturer's site to show more complex information.

The seats and tables disposed along the boulevards act as rest places where the customer can seat read carefully the information on the cart, order a coffee, and so on. In the future supermarket, shopping will not be a stressing activity, but a leisure activity. After all, if the customer has no time to waste, (s)he can, by sending the shopping list in advance, have the items separated and packed, ready to go in a drive-through fashion. The box container can be easily checked with her/his PDA acting as a RFID reader.

As final step, there are several proposals for fast check-out, like reading the cart by RFID, checking directly from electronic cart shop list, etc.

3.3 Use Case – Content-Centric Scenario

The use case presented here refers to a scene in the World Cup 2014 in Brazil.

Alice and Bob are enjoying moments of the Cup, and the personal devices (cell phones / PDA) will have a key role in this direction, with capacity for hundreds of Gbits, the integrated camera of 10 Mpx, video, high definition (HD), multiple flash memory cards and various radio access technologies (LTE 4G), WiFi (802.11n) and WiMAX (802.16e).

Alice's real-time updating her micro-blog with presence information, temperature, video and sound captures, and the pictures of her beside his sister into an unforgettable moment in their lives. Moreover, they are participating, with other spectators in the stadium, in a competition where the best images and videos are projected on big screens in the stadium and made available in real time over the Internet.

Bob is also participating in the interactive entertainment services provided by the organization via Wi-Fi access points. He has a permanent short-range connection with Alice and a mini 640 GB memory card in his wallet which also is part of his personal network. This personal network is federated to an Internet (cloud) data storage service. Data stored in high resolution on mobile devices are backed-up and replicated in a secure way via opportunistic communication with low cost connectivity (e.g. ad-hoc 802.11n access points).

Scenarios of Evolution for a Future Internet Architecture 69

The game finished and they are happy! Bob looks at his device, and notices a notification that a copy of the videos and images generated in the last 3 hours were stored successfully in the mini-HD which also maintains synchronization with the media servers (HDTV, computers) in his house.

It's time to celebrate, and friends decide to meet Carol, a friend of Alice, at her home. Each one is expressing her feelings about the goal that qualified Brazil for the final against Spain. Bob wants to show to his friends the video with the moment prior the goal, with a zoom on the image of the player who scored the goal. However, the battery of his mobile device is discharged and he forgot his wallet-size high-speed charger in the car. But he never forgets his personal identity device with biometric authentication system that allows him to access remotely information in any access point.

He uses Carol´s LCD TV 50 "´s multimedia panel browser that auto-configures Bob´s preferences, which were discharged at the service establishment. The friends watching the video decided to edit several scenes. The final piece is very good, and one of the friends asks to send a copy to his father who is traveling abroad Brazil. The multimedia device in Alice´s room, allows a cloud service video editing to create the video clip and to send the final file to the father of Bob´s friend, who has a good connection that supports high resolution format. The father receives an instant notification about the availability of content, accepts the reception, and within minutes he sends a new message of thanks including a photo of the little town in Finland where he is in a business travel.

4 Common Attributes of the Three Scenarios

The identification of attributes, characterizing each scenario (user-centric, object-centric and content-centric) is the next step toward the network requirements (see Figure 1). In this sense, we identified three common network´s attributes as key points to future Internet architecture: (L1) Mobility and Ubiquity, (L2) Capacity, Reliability and Availability, (L3) Security and Privacy.

4.1 Ubiquity and Mobility (L1)

Ubiquity and Mobility refer to anywhere/anytime attributes. Wherever/whenever a person (or an object) is, in the World, he/she/it should be able to communicate, without minding about boring jobs like "service availability", "set-up" or "new contract sign-in".

4.2 Capacity, Reliability and Availability (L2)

Once "the net service" is easily available, the second question is: "how well does it work?" Capacity, reliability and availability refer to this second question.

The diversity of applications (services) and the quantity of devices connected via ad hoc or fixed networks to Internet requires high capacity of the Internet architecture. The high capacity attribute in this context refers to deal with characteristics such available bandwidth, addressing space to identify billion of objects, power consumption save of devices and self-management and self-healing properties (autonomic networks). In this sense, capacity is directly related to the aspects of reliability and availability. Future Internet architecture must support mechanisms for survivability in the most diverse scenarios in both urban and rural areas. The reliability of connection in the infrastructure and the availability of the network are key requirements that can be implemented through the use of autonomic computing and intelligent algorithms that adapt the context to the real time changes in the applications. In this dynamic environment, with a high degree of mobility and pervasiveness, users are connected anywhere/anytime requiring efficiency and availability from the infrastructure network. The high degree of availability of connection will reflect directly on the user´s satisfaction.

4.3 Security and Privacy (L3)

The last but not least question is: as far as the net is easily reachable, and as far it works well, "how safe is it using it?"

Future Internet Architecture must include requirements for security and privacy that are already in use today as well as develop new and more comprehensive security systems. Due to the use of different communication technologies and different communication devices and applications, mechanisms to ensure the reliability of such communications should be incorporated and offered in various levels of security (users, devices and content´s authentication).

The techniques of digital signatures currently in development should be used to ensure authenticity, integrity and privacy of contents, users' preference data and different levels of interaction performed in mobile devices and in ad hoc networks. Regarding the content networks, mechanisms to ensure the copyright of the digital content should be used for authentication of copies. The use of globally-unique identifiers should enable uniquely the identification of data, objects, devices, sessions, applications, people, facilitating the creation of security mechanisms and authentication based on public key encryption to prevent DoS attacks.

Although these groups of the attributes are common to the three scenarios, each one has specific properties for each future Internet scenario, as follows.

4.4 Specific Attributes of Object-Centric Scenario

4.4.1 Ubiquity and Mobility

In terms of mobility, the objects' behavior may be as follows:

- Fixed, when they are installed in a fixed geographical point;
- Micro mobility, when the object's mobility is restricted to a confined area, for example, a robot that is always inside a house;
- Macro mobility, when the objects travel through diverse areas and contexts, e.g. when they are inside an automobile.

Scenarios of Evolution for a Future Internet Architecture

The last scenario is the most challenging because some extreme situations can occur, e.g. when sensors are collecting information inside a turbulent ocean.

The ubiquity concept for objects can change if there is or not a communication infrastructure where the objects are located. In affirmative case, the objects should be able to communicate with the diverse types of access networks. In negative case, the objects should participate of an ad-hoc communication network.

The problem of location of an object, in both conditions described above, presents two distinct challenges: (i) to locate the objects inside the network, i.e., to identify its virtual address; (ii) to identify its current physical location in emergence applications such as fire detection, rescue services, etc.

4.4.2 Capacity, Availability and Reliability

The various communicating objects have different traffic characteristics, such as:

a) **Transmission of short messages at regular time intervals**: this type of communication occurs when sensors are dispersed in wide geographical area. It is characterized by the transmission of short messages (some bytes) at intervals reasonably dispersed (e.g. one every ten seconds). However, total traffic can be high due to the large number of sensors involved.

b) **Transmission of video streams**: probably more critical than (a), the monitoring services (e.g. public places or residential monitoring services) generate high data traffic and are critic in terms of loss of information.

c) **Occasional transmission of high volume of data**: can occur, for example, in an emergency situation when a patient monitoring system needs to send data to an ambulance, or vice versa.

d) **Directionality**: in general, the traffic is unidirectional and upstream. However, some applications can have high bidirectional traffic, such as inter-vehicular communication or communication among objects in constellations of mobile sensors.

e) **Low round trip delay**: in applications involving process control (sensor-actuator), i.e., the delay time between sending the information captured by the sensor and the action command sent by the remote server should be low. Considering that the reaction time of a typical human is around 100 ms, round trip delays between 1 and 10 ms, are reasonable. It corresponds to frequencies between 100 Hz and 1 kHz (the vibration of an electric motor is 60 Hz and a typical automobile engine runs at 3,000 rpm).

Other challenges to a support network for communication between objects are:

- **Ability to prioritize and redirect critical messages:** as a part of messages between objects can refer to a critical traffic (security systems, medical emergency, drive commands, etc.), it is necessary that these messages can be sent with a higher priority even in case of network overload.

- **Low tolerance for delays**: generally speaking, humans are tolerant to delays in the sense that they are capable of take decisions in the event of an unexpectedly high delay. The machines do not have this capacity. Thus, the maximum time delay is a constraint to be considered, in both network design and machines design.

Another aspect related to this topic is the scalability, i.e., the possibility of inserting and removing objects, especially in ad hoc communications, maintaining the overall performance.

4.4.3 Availability

Ideally, a communication network must be available every time. However, in ad hoc communications, the network may be unavailable due to inactivity of the nodes. This situation can occurs, for example, when the objects are inside a car leaving or reaching certain areas and communicating via ad-hoc networks or when the network nodes are objects that can be turned off by humans.

4.4.4 Reliability

Ideally, when a human or an object sends a message in the network, he hopes the message reaches the destination. However, this situation does not always occur in a communication between objects and one of the following situations can occur:

- **Loss tolerance but low tolerance for delays**: in the world of communication between humans, this occurs mainly in telephone communication and transmission of images. It is possible that in some situation, automated machines or robots require images to be oriented, leading to this scenario of loss tolerance.
- **Tolerance to delays**: in some cases, communication between objects can tolerate some delay (of the order of seconds or minutes), such as the transmission of information on the temperature. In other cases, communication may allow a delay of hours, but the message needs to be delivered without fail, or can not be destroyed by time-out mechanisms of routers. This applies, for example, a sensor whose battery is ending sends an alert message to a server.
- **Intolerant to delays and losses**: in some cases, the message may be intolerant to delays and losses, as actuators commands.

Another situation regarding reliability refers to intermittent transmission, due to radio transmission. This is worsened when the object is in motion. Thus, the system must:

- Employ appropriate frequencies and protocols for a more robust transmission;
- Be able to self-recover quickly from situations of s loss of communication.

Scenarios of Evolution for a Future Internet Architecture 73

Another aspect of reliability is concerning with sensor networks including many nodes with low processing capacity. In this case is desirable to have a routing protocol able to perform an optimum routing balancing based on a set of parameters such as maximum delay, energy saving and data integrity.

4.4.5 Security and Privacy

The security of communication between objects has different aspects of communication between humans, such as:

- **Integrity of the message**: in general, objects have a little protection for abrupt break of the communication and corruption. When a human receives a false message, it is able to infer the existence of doubt about the proceeding and checks the information. Similarly, if communication is interrupted abruptly, the human is capable of making a new connection and takes actions to retrieve the message. However, the machines, in general, do not have such intelligence. Thus, in the communication between objects, message integrity is essential.

- **Misrepresentations**: various forms of attack refer to the misrepresentation, i.e. the attacker pretends to be a machine or person. Humans (sometimes) are able to identify this situation and take the appropriate action. The machines are more fragile in this sense. Thus, a support network for communication between machines must provide big robustness against this type of attack.

- **Denial of Service (DoS)**: another weakness point of the machines is its low or almost zero protection when a series of machines attack simultaneously sending messages to the target machine, overloading its processing. Ideally the network should provide reasonable protection against such attacks.

- **Different level of data protection**: the objects may contain information highly public, highly private, sensitive information that can be only accessed by authorized agents and private information, at opportune moments, but that may be publicized.

- **Physical integrity**: sensors that are remotely installed in dirt places and with high degree of vandalism. Thus, the partial or total damage to these devices may be a common occurrence.

Moreover, the device or machine itself, require physical protection mechanisms and the following cases must be taken in account:

- How to identify a sensor that is not communicating due to own problem, not on the network;
- How to locate a machine if it was stolen;
- How to infer information that should be filled by the sensor and
- How to create a "virtual sensor".

4.5 Specific Attributes of Content-Centric Scenario

4.5.1 Ubiquity and Mobility

The applications in a content network operate using identifiers of information's objects as basic primitives to access and manipulate the bits of information associated to these identifiers. In this process, the interface between applications and network should be agnostic to the access network technology, providing the connectivity and the location of the terminal in the network. The identifiers enable retrieval of information objects by the applications. Thus, it is possible to provide mobile content natively decoupling the identifiers of information from their physically location information in the network.

4.5.2 Capacity, Reliability and Availability

The Future of the Internet must provide mechanisms to address the growing of digital content stored and maintained in the network. The global communications infrastructure should support information management in larger scale.

Capacity: increase in the network due to reduction of redundant communications, use of caching mechanisms and other techniques for more efficient distribution of content (e.g. multicast, multipath, network coding). Ability to absorb spikes due to legitimate demand for a specific content and resistance to DoS attacks (security attribute).

Availability: increase availability of information´s objects in the network using identifiers and having a consistent distribution of exact copies on the network by multiple sources.

Reliability: improved reliability of the network as a platform for information distribution by the distributed storage of information and access to them, disconnected from the network topological information. The mechanisms that guarantee the authenticity of information help in making the network more reliable in relation to content that is available and that is delivered.

4.5.3 Security and Privacy

Mechanisms dealing with digital content protection, authentication and version control must be strongly supported as well as mechanisms that provide the network the ability to block unsolicited data by users. This last item can be implemented in both, the network layer and the application layer.

In a content-oriented model, the network natively behaves as a platform for distributing content. Another important factor is the ability to differentiate between a flash crowd behaviors done for thousands of computers infected with malware that perform requests to a particular site during a period of time. In this sense, the network must be able to detect this type of behavior and trigger the built-in procedures that guarantee the availability of services to users.

Scenarios of Evolution for a Future Internet Architecture 75

The content generated by producers must contain mechanisms for generating identifiers that allow them automatically to verify the integrity and authenticity of data received, and optionally, the identity of the actor (producer or intermediary) to guarantees the authorship of the information. This kind of security should not include information about the servers' locations or the communication channel available, but must be self-authenticate by the received content.

The design of a network focused on information brings new (or refined) types of properties such as authenticity, integrity, timeliness or reliability in the information processing chain. These items are considered fundamental properties that must be provided by new network architecture, especially one contemplating intermediaries that can perform substantial processing, storage and distribution from its own network. Beyond these basic properties, is considered essential to have a mechanism to validate the identity of users, software agents, and the purposes for which they are using the network.

The network should provide new security mechanisms to avoid more sophisticated security attacks. The ability to replicate data and to relieve them of massively location information along with new security mechanisms are intrinsic parts of the information that contribute to having a network more robust, secure and reliable.

The issues of data availability as a result of new mechanisms for storage and distribution of data and their reliability should be separated from questions related to copyrights of material subject to copyright.

4.6 Specific Attributes of User-Centric Scenario

4.6.1 Ubiquity and Mobility

This scenario requires the personal mobility, terminal portability of services and interfaces.

People are mobile by nature. Although mobile devices, smartphones, netbooks and so on already help a lot, the impression one has it that there is still a large room for improvements. People still have to carry heavy bags, deal with somehow not so friendly devices; must explicitly get (order) access to certain networks or services or, on the other hand, may be dangerously incurred on unauthorized services and charges. A virtual personal assistant should take on care such details like giving the right information a person needs, wherever he/she is, getting information about physical and virtual context.

This involves the concept of "Virtual Home Environment", where users can recreate their personal environments anywhere, anytime.

4.6.2 Capacity, Reliability and Availability

Regarding the network, this scenario involves a great capacity for distributed processing and large distributed databases. The availability of the network for the provisioning of personalized services and the QoS requested by the user in real time requires a large capacity network management and use metadata for service customization.

As services become more ubiquitous, pervasive and reliable, naturally their use increases and network capacity, reliability and availability become still tighter.

4.6.3 Security and Privacy

The user-centric scenario has as main feature the ability the user interact, build relationships and network with other people and even manage and control the network for its own use. This variety of new features aimed at the user must be implemented through mechanisms that provide a high degree of confidence in physical infrastructure and services. The separation of physical infrastructure provider and service provider can be an interesting alternative to separate the network functionality and facilitate the provision of new services and management of each layer.

The network should give the user the ability to be anonymous in relation to other users, vendors and institutions and to choose the desired level of security depending on the device they are using. Context-sensitive security mechanisms and solutions for authenticating users and their devices in order to minimize attacks should be also provided.

5 Conclusions

This chapter presented three future Internet evolution scenarios and three key common group of network attributes for them: (L1) ubiquity and mobility, (L2) capacity, reliability and availability and (L3) security and privacy. The specific properties for each scenario were also presented and may help in the next phase of the research to identify the network requirements for a future Internet network architecture.

Yet, although scenario building be a good technique to identify future needs and to set unforeseen requisites, it is both limited to our current vision about what the future can be (and it can be very different indeed) and, at some degree, to questions that we can solve with a short-range solutions. Therefore, one of efforts of the team is to do a continuous forecasting research, and check continuously the findings and new insights.

Acknowledgments. We would like to acknowledge funding provided by the FUNTEL (Brazilian Funding for Technological Development of the Telecommunications) of Ministry of Communications for this research and CPqD Foundation.

References

1. Akyldiz, I., et al.: A Survey on Sensor Networks. IEEE Communications Magazine 40(8), 102–114 (2002)
2. Braden, R., et al.: Developing a Next-Generation Internet Architecture, ISI white paper (2000), http://www.isi.edu/newarch/DOCUMENTS/WhitePaper.pdf (accessed April 2010)

3. Carzaniga, D., et al.: Content-based addressing and routing: A general model and its application (2000)
4. Carzaniga, D., Wolf, A.: Content-based networking: A new communication infrastructure (2001)
5. Creech, H., et al.: Mapping the Future of the Internet onto Global Scenarios – A preliminary view (2009), http://www.iisd.org
6. Ganapathy, S., Wolf, T.: Design of a network service architecture. In: Sixteenth IEEE International Conference on Computer Communications and Networks, ICCCN (2007)
7. Haggle: Project IST 027918: An Innovative Paradigm for Autonomic Opportunistic Communication - Deliverable 6.1: User-Centred Design of Haggle Applications (2006)
8. Heinzelman, W., et al.: Adaptive Protocols for Information Dissemination in Wireless Sensor Networks. In: Mobicom 1999, Seattle, pp. 174–185 (2009)
9. ITU-T Internet Report, The Internet of Things. 7th ed. Genebra (2005)
10. Jacobson, V.: If a clean slate is the solution what was the problem? Stanford "Clean Slate" Seminar (2006)
11. Jacobson, V., et al.: Networking Named Content. In: CoNEXT 2009, Rome, Italy (2009), http://www.parc.com/content/attachments/networking-named-content-preprint2.pdf (accessed April 2010)
12. Kahn, J., et al.: Emerging Challenges: Mobile Networking for "Smart Dust" (2000); Koponen, T. et al.: A data-oriented (and beyond) network architecture. SIGCOMM Comput. Commun. Rev. 37(4), 181–192 (2007), http://www-ee.stanford.edu/~jmk/pubs/jcn.00.pdf
13. Koponen, T., Chawla, M., Chun, B., Ermolinskiy, A., Kim, K., Shenker, S., Stoica, I.: A data-oriented (and beyond) network architecture. SIGCOMM Comput. Commun. Rev. 37(4), 181–192 (2007)
14. Trossen, D. (ed.): PSIRP Conceptual Architecture of PSIRP Including Subcomponent Descriptions (D2.2) (June 2008), http://psirp.org/publications
15. Roberts, J.: The clean-slate approach to future Internet Design: a survey of research intitiatives. Ann. telecommun. 64, 271–276 (2009)
16. Rothenberg, C., et al.: Towards a new generation of information-oriented internetworking architectures. In: ACM CoNext, First Workshop on Re-Architecting the Internet (Re-Arch 2008), Madrid, Spain (2008)
17. RWI 2009, Real World Internet Position Paper (2009), http://rwi.future-internet.eu/images/c/c3/Real_World_Internet_Position_Paper_vFINAL.pdf (accessed April 2010)
18. Santucci, G.: From Internet of Data to Internet of Things. In: International Conference on Future Trends of the Internet (2009), http://ec.europa.eu/information_society/policy/rfid/documents/Iotconferencespeech012009.pdf (accessed April 2010)
19. SAP, The Internet of Things: Reality or Hype? (2009), http://www.sap.com/about/company/research/irf/2009/index.epx (accessed April 2010)
20. UCM, User Centric Future Media Internet - white paper (2008), ftp://ftp.cordis.europa.eu/pub/fp7/ict/docs/netmedia/ucm-white-paper_en.pdf (accessed April 2010)

Future Network Architectures: Technological Challenges and Trends

Antônio Marcos Alberti

Instituto Nacional de Telecomunicações (INATEL), Av. João de Camargo 510,
Santa Rita do Sapucaí 37540-000, Minas Gerais, Brazil
antonioalberti@gmail.com

Abstract. The Internet became an extraordinary artifact for information exchanging, and it is now being considered a strategic infrastructure even for countries development. However, its success has fueled an upward spiral of new applications, which require increasingly intensive use of its capabilities, putting a lot of pressure on a design originally drawn for a scenario remarkably different from the current one. Concerned with actual Internet evolution and its limitations towards a future global information infrastructure, many research initiatives started to reinvent or rethink Internet, so it can fully assume the role we are assigning to it. This text discusses technological requirements, challenges and trends towards future network architectures. It covers latest new Internet design approaches and their main innovations.

1 Introduction

The Internet can already be considered one of the most impressive human artifacts. It became an essential part of our society, economy and institutions, and it is profoundly changing the way we exchange information. Virtually almost everybody is directly or indirectly affected by the available Internet infrastructure. Nobody doubts about the importance of the Internet for the evolution of our information society. However, despite its tremendous success, the Internet was originally designed decades ago to interconnect computers and applications. Nowadays, we want to use it as a global information infrastructure in order to meet a vast diversity of uses and aspirations. This shift in the original usage plus decades of incremental changes and the difficulty to keep tracking evolution began to awake the need for change. Network researchers and even prominent Internet designers (e.g. Van Jacobson, David Clark) started to ask if current layered HTTP/TCP/IP[1] Internet is ready to be such global information infrastructure as well as to support the tremendous increasing of new applications traffic, the drastically increasing in the number of nodes, new 3D and immersive applications,

---

[1] IP, TCP and HTTP stand respectively for Internet Protocol, Transmission Control Protocol and Hypertext Transfer Protocol [1].

rich e-media and content distribution, and so forth. The Internet as well as IP convergent networks limitations and uncertainties motivated worldwide efforts for a new Internet design [2]. Many researchers started to advocate for a clean-slate design for the Internet. According to Stanford University Clean Slate project the goal of clean-slate design is to "reinvent the Internet to overcome fundamental architecture limitations, to in-corporate new technologies, to enable new classes of applications and services and to continue to be a platform for innovations and thus be an engine for economic growth and prosperity for the society" [3][4]. According to Future Internet Design (FIND) initiative[2] the aim is to "invite the research community to consider what the requirements should be for a global network of 15 years from now, and how we could build such a network if we are not constrained by the current Internet – If we could design it from scratch" [5][6]. The idea is to make an unconstrained redesign of the Internet, using what we learn so far.

Although there are some clean slate approaches worldwide [7], other proposals seek for innovation through evolutionary means, trying to create mechanisms to continue the evolution over current Internet infrastructure. Others, however, advocate the idea that new network architectures will be a composition of both evolutionary and clean-slate approaches. Regardless of the merit of which vision promises better results, the idea of reinvent or rethink Internet has a lot of open issues, which constitute a new and exciting research horizon. In this scenario, the aim of this text is to address several technological questions behind approaches for future networks architectures. For example, which are the requirements for a new Internet architecture? Are the requirements inter-related and/or inter-dependent? Which are the main challenges for design? What about conflicting aspects? What technologies could be used? What are the technological trends? These questions illustrate the scope of this text. Obviously, the intention here is to provide insights on these questions, rather than answer them. Another observation is that nobody has a "crystal ball" to look into and see what will happen in the future with a 100% certainty. Therefore, the trends presented here are based on current technological trajectory, and thus can undergo significant changes in the near future. The literature has some examples of misquotations, which illustrate how risky is trying to predict technological trends. A famous one is "There is no reason anyone would want a computer in their home" from Ken Olsen, founder of Digital Equipment, in 1977 [8]. Therefore, this text seeks to show just what would be the "next point on a graphic" according to the current level of consensus in the research community. This is what we mean by a "trend" in this text.

To explore the diverse world of possibilities behind future networks design, the strategy used here was to select a group of technological requirements and discuss each one based on latest proposals and their innovations. Of course, this work is multidisciplinary, since Internet is relevant to our entire society. However, our adopted approach was to limit the question to information and communication point of view. Despite of this limitation, some presented references look further to other knowledgement areas.

---

[2] Long-term initiative of the United States of America (USA) National Science Foundation (NSF) Networking Technology and Systems (NeTS) research program.

The remainder of the text is organized as follows: Section 2 presents technological requirements, challenges and trends in processing, storage and connectivity infrastructures. It encloses wired and wireless transmission medium, including optical networks, cognitive radio and wireless sensor networks. Network virtualization is also discussed in the context of wireless networks. Section 3 focuses on network entities identification, mobility and localization. Section 4 provide insights related to adaptive networks, manageability, network self-awareness and self-emergent behavior and their application to control and management. Section 5 covers aspects regarding information context, identity, naming and routing. Section 6 aims to discuss security, privacy and trust in future network architectures. Section 7 focuses on services and applications, including architecture neutrality and flexibility. Finally, Section 8 finishes the text covering design simplicity, evolvability, and sustainability.

2 Capacity, Efficiency, Performance, Ubiquity, Scalability and Generality

Technology evolution in terms of digital storage and processing has remained consistent in previous decades [9]. Storage capacity has advanced significantly with solid state flash memories and high capacity magnetic disks, while prices are decreasing enormously. Solid state flash memories consume a small fraction of magnetic disks energy and are becoming more and more overspread, from pen-drives to a replacement of hard drives. Other types of memories, such as polymer memories, could increase storage capacity even more. Processing capacity also continues to grow, while chips energy consumption is reducing. Such effort is essential especially for mobile devices and sensors nodes, where severe energy constraints exist. The objective is to reduce battery energy drain as well as heat dissipation. The research for alternative energy sources remains, such as dye-sensitized solar cells for mobile devices [10]. High-performance computing based on supercomputers or computer clusters is achieving teraflops and evolution proceeds to petaflops in the next decade [9]. Moore's law continues to apply for processing and memory capacities as well as for digital displays. Graphene transistors could replace current silicon ones, creating ultrafast processors [11]. Display technology has advanced tremendously in later years, allowing better quality and larger screens, substantially improving the quality of experience and allowing new forms of digital interactivity. Printed electronics could also contribute to reduce production costs [12][9]. Other promising technologies are silicon photonics [13], Micro-Electro-Mechanical Systems (MEMS), nanotechnology, Nanoelectromechanical Systems (NEMS), quantum computing and communications, carbon nanotubes, etc. Besides a vigorous debate about how long processing, storage and display technologies will continue to follow Moore's law, there is no doubt that these technologies deeply affected and will keep affecting Information and Communication Technologies (ICT) industry as well as people's life.

Internet was designed in an era were computing processing and storage capacities were severely deficient and related costs extremely high. Of course, such fact impacted on design possibilities and Internet deployment for years. For example, IP addressing space designed to connect thousand of networks and hosts [14], since routers had too few capacity to deal with bigger datagram's headers. Such reality was used to derive some of principles that guided Internet design. They began to be evaluated in scale after ARPANET[3] and NSFNET[4] interconnection and the adoption of the TCP/IP protocols suite [1]. The evolution continued, increasing the number of networks and hosts, always taking advantage of emergent processing and storage technologies and their costs decrease. In July 2008, the extraordinary and even unpredictable amount of 570 million hosts were connected to the Internet, and many believe that growth will continue to reach 3 billion in 2011 [15]. This growth will be in part motivated by IP based fixed/mobile convergence, which will bring millions of mobile devices to the network [15][16][17]. In some countries, mobile devices have outnumbered PCs on Internet [14][15] and there is more to come. According to Saracco [9], in "2005 only a tiny fraction of microprocessors produced ended up in something that could be called a computer." This shows that more and more devices are becoming computationally capable and sooner or latter will be connected to the Internet.

Heterogeneity of devices will prevail, from supercomputers to nanotechnology devices. The abundance of processing and storage could enable that virtually any artifact, from clothing to buildings, to be connected to the Internet as Network Enabled Devices (NEDs) [17]. Such trend is being generically termed Internet of Things (IoT) and the proposal to integrate real world into the Internet calls Real World Internet (RWI) in the context of European Union Future Internet Assembly (FIA) activities [17]. Japanese New Generation Network (NwGN) Project Akari[5] [2] determines the relation of real-world aspects of your society to entities in the NwGN space as the reality connection design concept. Cheap computing is leading us toward the paradigm of ubiquitous computing, which allows the creation of smart environments or ambient intelligence. It is the "age of calm technology, when technology recedes into the background of our lives" as defined per Weiser [18]. There is considerable consensus that the amount of NEDs plugged in the Internet could reach the number of billions or even trillions [2][15][19]. Therefore, they would become the majority of connected devices in the Internet.

IoT and RWI could push the shortage of IPv4 addresses. Many are the predictions of the year in what the limit of available addresses in IPv4 will occur [20]. Current Internet numbers illustrate how the initial estimates for hosts, networks and autonomous systems were surprised by its success. Even in the 80's, the vertiginous progress on Internet number of hosts has drawn attention to the exhaustion of IPv4 address space. Nobody knows exactly when the limit will reach, but everything indicates that it is to come. The shortage of IPv4 addresses has resulted

---

[3] ARPANET stands for Advanced Research Projects Agency Network. It was created by the Defense Advanced Research Projects Agency (DARPA) of the United States Department of Defense.

[4] NSFNET stands for National Science Foundation Network.

[5] Akari project is promoted by Network Architecture Group of New Generation Network Research Center in National Institute of Information and Communications Technology (NICT), Japan.

in several technologies to deal with the problem, such as Classless Inter-Domain Routing (CIDR), Network Address Translation (NAT), Dynamic Host Configuration Protocol (DHCP) and IPv6 [1]. CIDR creates a binary level division in the host and network portions of IP addresses, allowing better utilization of the address space. DHCP attributes dynamically addresses to devices, while NAT allows the use of virtual address spaces inside autonomous systems. NAT is generally considered responsible for the loss of transparency in the Internet [2], since traceability is compromised inside autonomous systems. IPv6 vastly increased address space of its predecessor IPv4 using 128 bits addresses, thus allowing up to 3.4×10^{38} hosts. IoT can accelerate IPv4 address space exhaustion, while IPv6 alleviates the problem, but still generates concerns, regarding other aspects, such as routing, scalability, QoS, security, etc [15][2].

According to Minnesota Internet Traffic Studies (MINTS) the annual Internet traffic growth rates were about 50-60% in 2008 and about 40-50% in 2009. This means that Internet traffic could grow roughly 30-100 times in the next decade. Also, according to this reference, year-end 2009 monthly Internet traffic estimate was circa 7,500-12,000 petabytes, i.e. $7.5-12 \times 10^{18}$ bytes or exabytes. To give an idea, this traffic is equal to about 300 million single layer blu-ray^{TM} discs content monthly. Akari project aggressively estimates that traffic could increase 1.7 times per year in Japan in the next years, producing an expansion of 1000 times in 13 years [2]. If we extend such numbers to present 10 Mbps fixed access networks, the required data bit rate would be 10 Gbit/s in 13 years. A candidate technology to meet this requirement is Fiber-To-The-Home (FTTH) [2][15][9]. In the backbone, current 1 Tbit/s nodes would require 1 Pbit/s, i.e. 10^{15} bits per second. European Union Future Internet Assembly (FIA) cluster Management and Service-aware Networking Architectures (MANA)[16] defends that "in the core we are moving from gigabit networks to terabit/s."

2.1 Optical Networking

To attain such capacity, not only state-of-art optical transmission and switching will be required [2][15][7], but also deploying more fiber [9]. Ultra-Dense Wavelength Division Multiplexing (UDWDM) and Optical Time Division Multiplexing (OTDM) are candidate technologies [2]. UDWDM systems will allow thousands of wavelengths in one single fiber. Ultra-high speed OTDM systems of 160 Gbit/s – 1 Tbit/s have already been tested, respectively on [21] and [22]. Another tendency is the so called all-optical networks. Some devices that today still use electronics should be transformed into fully optical, e.g. wavelength converters, multiplexers, regenerators, transponders, switches, cross-connects, etc. For others, however, we need more research. This is the case, for example, of optical buffers for Optical Packet Switching (OPS) technology. Fiber Delay Lines (FDL) provide temporarily "storage" for light as it propagates over a fiber loop. However, due to the speed of light, it is expensive to construct loops with more than a few microseconds of storage as well as large optical crossbars to manage storage. Other challenges include optical packets contention resolution and loss. Therefore, the search for an optical buffer analogous to modern electronic ones continues.

Optical packet switching also needs extremely high switching times. MEMS and NEMS are promising technologies for the construction of extremely fast all optical switches. New fiber optics, such as Photonic-Crystal Fiber (PCF) [23] and Photonic Bandgap Fiber (PBF) [24] have a crucial role in creating ultra-wideband systems.

PCFs are specially designed fibers that have unique properties due to the planned distribution of air holes along their entire length. A large range of properties variations could be obtained by varying fiber design. Unusual chromatic dispersion of PCF can be used to generate supercontinuum spectrum [25][26], i.e. to broad light spectrum. Another recent optical technology is Optical Frequency Comb (OFC) [27]. It is a series of spectral impulses exactly equidistant generated by mode-locked femtosecond lasers. Temporal coherence property of PCFs can be explored to generate optical frequency comb [27][28]. OFC technology enables development of extremely precise optical synchronization. Ultra-wideband optical transmission systems could be created combining both techniques [2]. In 2006, Miyagawa et al. demonstrated an ultra-dense 10000 wavelengths 2.5 GHz-spaced WDM system that uses such combination of technologies [29].

Finally, in the visible border of ICT future are the quantum technologies. Quantum Key Distribution (QKD) enables safely sharing of confidential keys [30]. QKD experiments have been made not only over fiber optics [31], but also over unguided medium [32]. According to Masahide [33], "quantum cryptography has reached its final stage moving through practical application". Quantum teleportation is the technology for transferring of quantum information from a source quantum system to a destination one. Information transfer occurs by means of entangled photon pairs. A provocative question is: do the information teleportation happen faster than the speed of light? [34]. Quantum teleportation have already been deployed in experimental testbeds [35][36][32]. It is a fundamental part of quantum communication systems, which are considered a promise to overcome Pbit/s limits in optical communications [33]. Such results shown that quantum networks are becoming feasible.

2.2 Wireless Networking

What are the requirements of a future wireless environment? Ubiquitous connectivity, devices heterogeneity, high capacity, poor spectrum usage, high efficiency, low interference, low energy consumption, mobility, flexibility and autonomous operation are the main concerns. Ubiquitous computing leads to ubiquitous connectivity, i.e. connectivity anywhere, anytime, in anyplace, to anyone. Such connectivity could be provided by centralized approaches were cells size range from very small (picocells and femtocells) for high bit rate dense areas, up to large cells to cover rural areas. Nevertheless, decentralized mesh networks are emerging as strong candidates for future wireless connectivity [9]. Independently, the aggregated traffic will be carried mainly by high capacity optical networks [9]. Global coverage will be necessary and can be achieved by integrating terrestrial and satellite networks [15]. Another fundamental requirement is to stay connected while moving, i.e. mobile connectivity [37]. Also, it will be required to deal with a large variety of devices, with different constraints, e.g. sensors, cars, mobile phones, PDAs, etc. New high capacity radio technologies will be necessary to meet predicted traffic growth. Going back to the Akari Project traffic increase rate estimative (1000 times in 13 years), current mobile devices access rate of 1 Mbit/s would need to be expanded to 1 Gbit/s. Interestingly, this rate also appears as the required one for short-range wireless communications in European Union EIFFEL[6] Think-Thank initiative [14] as well as in the scope of 7th Framework Programme (FP7) OMEGA project. One technology that OMEGA project considered achieving such rate is wireless optics in infrared wavelengths. Indoor coverage is addressed by line of sight links, while eye safety requirements are considered. Such solution is also attractive because it escapes from the ultra congested Radio Frequency (RF) spectrum. Nevertheless, RF technologies like Multiple-Input and Multiple-Output (MIMO), cooperative communications [38], Ultra-Wide Band (UWB), Orthogonal Frequency Division Multiplexing (OFDM), digital processing, antennas and coding are pushing data rates near Shannon theoretical limit [9]. This proximity to the theoretical limit indicates that the evolution of wireless communications is demanding research efforts not only on physical portion, but also in areas that could ensure better utilization of network and spectrum resources.

2.2.1 Cognitive Radio Networks

It is a common sense that more efficient wireless spectrum usage approaches must be achieved, since a considerable part of the spectrum is underutilized today [37]. Radio spectrum is a key limited resource for future networks and its usage demands innovative, efficient and dynamic solutions. A multidisciplinary effort is required, since wireless spectrum is considered a natural (and public) resource and, therefore, is subject to national laws and regulations. Cognitive Radio (CR) appeared as an option to increase radio spectrum usage by means of dynamic spectrum allocation [39][40][41][2]. The term was coined by Mitola III in 1999 [39]. The idea is to share opportunistically the same radio spectrum between primary and secondary network operators. A primary operator is a licensed one, e.g. a cellular telephony operator, a television operator, etc. A secondary operator can be authorized to explore dynamically unused frequency bands (or frequency holes) assigned to primary operators.

Cognitive radio is based on the concept of Software Defined Radio (SDR), where physical layer signal processing is software-based instead of using traditional hardware and radio parameters are dynamically configured by software. SDR goal was also proposed by Mitola III [42], but a few years before (1995). For example, reconfigurable radio hardware could have implemented two different coding algorithms. It is the software that chooses the most appropriate one and reconfigures hardware architecture accordingly. In other words, one can say that the "intelligence" of the operation is left to the software. Operational decisions are taken according to the state of the radio environment as well as the physical hardware capabilities, such as antennas, digital signal processing, baseband varieties,

---

[6] EIFFEL stands for Evolved Internet Future for European Leadership.

modulations, codings, etc. This paradigm poses formidable challenges for radio designers.

Cognitive radios should be designed to achieve goals not only for the physical portion of the network (e.g., reduce interference with the primary networks, improve bit rate), but also for the upper portions (such as QoS, optimal routing, services needs, etc.). Thus, Cognitive Radio Networks (CRNs) are not only radio environment-aware, but also regulatory-aware, i.e. cognitive radio network must be aware of current radio environment in order to detect spectrum opportunities as well as primary operator activities. Also, CRN must be aware of current regulatory aspects, such as spectrum licensing for primary operators, spectrum share policies, in order to operate in accordance with approved regulations and licensing.

Cognitive radios must search constantly for new bandwidth opportunities, according to their hardware capabilities. When a spectrum opportunity is found, CR must decide to use it or not. In addition, the radios should process the received signals to identify primary users' transmissions. In case of detecting a primary signal, cognitive radios should stop their transmission immediately to do not interfere in primary signal. Shadowing areas, multipath fading and other phenomena can cause false frequency opportunities detection, generating interference with primary users. Cooperation among CR can help to avoid interference problems [43]. CRN can experience highly disruptive physical links due to the traffic characteristics of a diversity of primary networks and the possibilities of different operating bandwidths. Thus, cognitive radio network topology varies according to opportunistic links availability. This could create non favorable scenarios for CRN traffic. A proposal to deal with such limitation is to allocate more than one frequency band for each logical link, allowing orthogonal simultaneous transmissions. Thus, the loss of a frequency hole implies on the reduction of a logical link transmission rate, instead of total disruption.

Cognitive radio networks require a lot of autonomous operation to achieve their goals [41]. As defined by Haykin [40], cognitive radio "uses the methodology of understanding-by-building to learn from the environment and adapt its internal states to statistical variations in the incoming RF stimuli by making corresponding changes in certain operating parameters in real-time". To provide efficient, reliable, and ubiquitous communication the radio must learn from the environment. It must cognitively adapt its operational parameters in real time based on Radio Frequency (RF) channel conditions. Therefore, a CRN needs self-organization mechanisms to perceive radio environment, to establish network links, to achieve cooperation among radios, to keep track of historical decisions on spectrum holes and interference, etc. Haykin argues in this direction [41]. Also, CR needs self-configuration mechanisms to autonomously configure physical parameter profiles. Like many other autonomous systems today, CRN operation could be assisted by a human network operator. The appropriate value for each physical parameter can be set up by self-optimization algorithms, minimizing negative aspects, i.e. interference, loss, resource unfairness, etc.; and maximizing positive ones, i.e. throughput, spectrum usage, availability, reliability, spectral efficiency, routing performance, etc. Also, transmission power must be optimized to allow coexistence. Self-control is required to distribute fairly network resources as well as to provide service-awareness, context-awareness, etc. Scheduling could be performed considering increasing situation awareness. Finally, self-protection allows CRNs to protect against attacks. We will return to autonomic network operation on Section 4.

An indispensable multidisciplinary challenge that appears in CRNs is the deployment of multidomain, multioperator environments, to avoid several stakeholders to turn on simultaneously and chaotically their CRNs. Such a scenario can bring prejudice rather than gains, i.e. poor and unfair spectrum usage. Therefore, some common framework will be necessary to provide fair and efficient spectrum usage among stakeholders, while maintaining healthy competition. Interestingly, such scenario appears to repeat itself in an operator network, where users' terminals compete each other for the available resources, while collaborate to exchange measurements and available resources. Clearly, there is a dispute of interests between cooperation and competition in CRNs, very similar to what happens to social animal species in the African savannah. They need to cooperate in order to achieve their goals, but there is a close race to obtained results. Social etiquette maintains the equilibrium. You should be thinking: is there any relation to the famous John Nash game theory? Of course, it is. Game theory is widely used to address spectrum sharing in CRNs. Haykin analyzed CRNs spectrum sharing as a game in 2005 [40]. More recently, he suggests a decentralized dynamic spectrum management approach, based on bio-inspired self-organization mechanisms [41]. Other proposals on spectrum sharing are based on optimization, stochastic and information theory techniques [44]. See Alkydiz et al. [45] for a survey on CRN spectrum management.

Beyond technical point of view, multidomain multioperator CRNs will demand some type of spectrum market, since primary operators invested a lot in licenses worldwide. Hourcade et al. [37] suggest the creation of a "real-time location-sensitive spectrum trading market" as a solution. Note that current spectrum allocation is extremely strict. Cognitive radio brings the possibility that secondary operators use frequency holes of primary licensed operators. In essence, secondary operators become a second level spectrum allocation. In the future, frameworks for Dynamic Spectrum Allocation (DSA) can go beyond two levels, creating different spectrum access options, increasing licensing models diversity. However, co-existence with minimum interference will continue being a requirement.

Cognitive radio networks are pushing new Internet architecture towards more dynamic, efficient, autonomous, self-aware and ubiquitous wireless networks. Indeed, CR is becoming a general research term for radio technology [46] and its influence and legacy is already presenting in new Internet architecture proposals [2][15]. CRNs research is by now addressing essential requirements of new Internet architectures such as fairness[7] competition, quality, efficiency, mobility, autonomy, semantics, context, etc. However, a more holistic approach is required to align CRNs research with post-IP proposals, integrating cognitive radio,

---

[7] In essence, Quality of Service (QoS) mechanisms can be seen as mechanisms to provide fair resource usage among traffic flows in order to achieve desired quality levels as established by negotiated traffic contracts. The term fairness reflects the need for balance between quality and resource usage.

wireless sensors, actuators and other NEDs in a truly convergent approach for future architectures. Imagine the possibilities of a cognitive radio enabled real-world Internet. Integration with optical networks is also required.

2.2.2 Real World Internet

Let's return now to the Internet of Things and discuss how it impacts on the already congested RF spectrum. The IoT concept is frequently attributed to the Auto-ID center of Massachusetts Institute of Technology (MIT) [47]. Examples of objects or things are household appliances, surveillance equipment, security equipment, sensors and actuators, bottles of wine, goods in a supermarket, etc. The identification of objects can be made using Radio Frequency Identification (RFID) or Near Field Communications (NFC). RFID is already used to prevent the theft of goods in stores, to provide traceability of products in supermarkets, and a million of other applications [48]. RFID systems are composed by interrogators and tagged wireless devices. An interrogator emits a radio frequency signal that triggers a tag to transmit backwards some pre-programmed information, e.g. identifiers. NFC is a technology for proximity wireless high frequency communication that enables devices to exchange information in short distances. NFC technology could be used to substitute traditional cards, tickets, identities and even money. "NFC phones are becoming our new electronic wallet" according to Belpaire [48]. Sensor nodes are devices used to monitor physical phenomena, such as rain, fire, pressure, movement, humidity, temperature, touch, etc. They have processing and communication capabilities. In the other side, actuators are devices that act over physical environment, such as pneumatic, electric and hydraulic actuators. IoT includes wired and Wireless Sensor and Actuator Networks (WSAN).

Such technologies could bring freshly and context aware physical world information to enrich virtual world applications, changing the way we perceive real world today [49]. WSN applications examples are monitoring of people healthcare, farmlands, manufacturing processes, industrial plants, houses, cars, disasters, traffic, transportation, environment and climate. Eventually, if sensors are placed on everything we know and the information collected are adequately provided for innovative applications, there will be a monumental growth not only in the number of devices, but also in the amount of information produced as well as in the traffic generated. RFID, NFC and WSAN technologies provide cost-effective means to interact with real world, from objects identification to context-aware information for innovative applications [15][19]. Using these technologies not only real-world information could be automatically captured, but also digital world could change the real world through actuators. Information collected by these technologies will be precious for new Internet services and applications. New Machine-to-Machine (M2M) applications could take advantage of fresh context-aware information.

Although these technologies are fundamental for RWI scenario, they bring formidable challenges to new Internet designers, mainly because they have several fundamental physical restrictions [50]. In general, wireless sensor nodes use batteries to power up sensing, processing and communication hardware. Therefore, energy consumption is a problem. When a sensor node depletes its battery it becomes inactive, and often, the batteries can not be replaced. Also, WSN is by nature a cooperative network, where nodes help each other to route information to a sink device or to process aggregated information. Therefore, if a node depletes its energy source it reduces network availability or impairs the function for which the sensor network was built. Also, sensor nodes spent energy processing information as well as doing storage, although it is communication that spends more energy [49]. WSNs have variable topology due to node failure, mobility or power variations. Also, WSN could be formed through sensor nodes dropped from a plane or other transportation over the interest environment, thus requiring self-organization, self-configuring and other autonomous functionalities [2][49]. Multiple access and routing need also energy-aware approaches, because collision spend energy unnecessarily and routing could consume nodes' energy unequally. Information aggregation, when possible, could reduce frequently transmissions. Depending on application, sensor nodes density can vary from very small to exceptionally large quantities [50]. Therefore, WSN availability increases as sensor nodes density increases. WSN is also vulnerable to a great diversity of attacks, from radio signal jammer, to sinkhole and battery drain attacks [51]. In addition, WSN could carry privacy sensible information, such as location, identity and other contextualized information. Other security aspects are node authentication, node authorization, data integrity, confidentiality and freshness [51]. RFID and NFC also have implications on privacy, security, localization, identification, etc. In summary, NEDs and their networks have a strong impact on several requirements of future network architectures.

To help realizing how RWI could deeply impact on a new Internet architecture, consider the following questions: How to deal with such tremendous number of devices? How to identify/address them globally? How to locate them physically? How to be sure that a NED is an authorized device? How to secure and protect NEDs privacy of information? How to trust in a certain NED and on its information? How much capacity is required to transport RWI traffic? How to protect RWI from attacks? How to answer previous questions considering the set of devices restrictions: energy, power, spectrum, etc. These questions are just to give us an inkling of the challenges behind RWI design. NEDs traffic will impact future networks capacity. Although, sensor nodes, RFID and NFC devices do not generate too much traffic individually, the aggregated RWI traffic can be representative.

Traditional security techniques could expend too much energy on NEDs, while inadequate security mechanisms could reduce devices lifetime, affecting network availability [51][52]. So, RWI needs energy-aware security approaches. Trust mechanisms appear to be useful to improve security in RWI, since information aggregation, correlation, filtering and routing could be based on trusted paths [53]. The collected data needs to be contextualized to allow delivering of the right information, to the right person or machine, at the right time [54]. Metadata could help to provide adequate data description based on ontology [55]. Support for devices mobility is also a requirement. Autonomous operation is required not only to organize and configure the network, but also to increase efficiency and life time. Self-management, self-healing, and self-optimization are also challenging aspects to be addressed towards operation complexity reduction (see Section 4). Plug-and-play capabilities would help also [19].

A diversity of information communication models would be supported, such as Peer-to-Peer (P2P), M2M, intermediate sink-sources [56], push-and-process [15], publish/subscribe [15][57][58], disruptive communication [2][19][17], event-based [59], cluster [9], sticker [9], contextualized [9], etc. Mechanisms to handle the large amount of collected information will be required as well as mechanisms to prune no longer desired information, e.g. mechanisms analogous to human memory cleaner. Also, some infrastructure will be required to expose adequately WSN capabilities to service development frameworks, taking advantage of available context [17]. Additionally, devices discovery mechanisms could be provided to address service level functionalities. For example, a certain service could provide the geographically closest temperature for web sites. This service would use a device discovery mechanism to locate a temperature sensor in some WSN and query it. Network embedded physical objects could have identities, communicate to people, machines or other network embedded physical objects, becoming aware of their environment and even presenting some form of "virtual personality" [17][9]. In this scenario, not only devices global unique identities, but also identity management mechanisms would be necessary [17]. See Section 3 for more details about identification and localization.

2.3 Providing Generality with Virtualization

Until now we discussed some technologies that provide the basic features for processing, storage and transport of information. Obviously, the list of technologies that have some importance in this substrate infrastructure does not stop here. A question that arises in this context is how to make this diversity of resources transparently and uniformly available to build future network architectures? In other words, is it possible to design a generic framework to expose all these resources simultaneously? The key word is generality. Without it, we are probably bound to create systems unnecessarily complex and inefficient, with high Operational Expenditure (OPEX). This is the case today. Telecommunications services are fully linked to existing transport technologies, making the creation of new services unnecessarily difficult. There is always some peculiarity that generates additional complexity as well as non-compliant aspects that lead to sacrificing efficiency. A solution could be a common framework for all substrate resources, enabling to expose them properly, regardless of their diversity. Such generic, common, uniform, transparent framework could be the base for future network architectures. On top of this infrastructure, we can build everything else, from networks to applications. Not surprisingly, there is already a technology that is practically a consensus for this purpose: network virtualization. Nevertheless, it is quite probable that other approaches to provide architectural generality will appear sooner or latter.

The origin of the term virtualization was the emergence of virtual machines in the 60's. A virtual machine is a machine made in software. In this context, virtualization can be defined as a technique that "introduces a software abstraction layer between the hardware and the operating system" [60]. This abstraction layer hides and homogenizes hardware computational resources to allow one or more operational systems to run concurrently on the same host computer, creating one or more virtual machines. According to Marinescu and Kröger [60], this abstraction layer was called Virtual Machine Monitor (VMM or hypervisor). The first hypervisor was conceived in 1967, in the scope of IBM's one-off research system CP-40. In 1972, IBM launched its virtual machine Operating System (OS) VM/370. The hypervisor can run directly over a host computer hardware (Type 1) or over a host machine OS (Type 2). Type 1 hypervisors are also called "native" or "bare-metal" hypervisors since its performance overhead is kept to a minimal. They control access to hardware resources providing isolation among OSs. Type 2 hypervisors are also called "hosted" hypervisors because they run as an application over a host OS. They provide virtual hardware resources to parallel guest OSs.

For many years, mainframe virtualization was essential to computing. However, with the PCs advent and the rise of multitasking operating systems, VMMs were left out in the 80's [60]. Today we are experiencing a strong return to hardware virtualization, driven mainly by OPEX and energy consumption reduction as well as improved hardware utilization. Hardware virtualization is also becoming popular at desktop PCs, allowing users to use several OSs. Such virtualization movement is affecting almost all areas of computing, ranging from machines, operational systems, storage servers, memories to mobile devices [61]. Even giant server farms and data centers are being virtualized under the name of cloud computing. It seems quite natural that virtualization also reaches the network.

2.3.1 Network Virtualization

Network virtualization aims to provide a high-level abstraction of network hardware [62]. It allows the creation of customizable, programmable and independent virtual networks [63]. Like other types of virtualization, network virtualization is not new at all. More limited versions of the concept have been used to create tunnels, Virtual Private Networks (VPNs) [63] and Virtual LANs (VLANs) in practically all OSI[8] layers. In 1998, X-bone research project received DARPA funding to deploy and manage overlay networks over the Internet [64][65]. Project results were used to implement Virtual Internets over the Internet through the use of a virtual link and network layers. The idea was to support the concurrency of multiple virtual Internets. More recently, network virtualization has been proposed to overcome traditional testbed limitations [62][66]. It was the emergence of virtual testbeds. Since network virtualization can allow diverse Virtual Networks (VNs) to share the same physical Substrate Network (SN) [61], diverse network slices can be created to allow simultaneous experimentation. In 2002, not only X-bone extended their work (X-Tend) to deal with virtual testbeds, but also Peterson et al. proposed PlanetLab [66], which became one of the most widespread research networks worldwide[9]. Also in 2002, White et al. [67] proposed Emulab, which uses overlays to allow distributed network experimentation. PlanetLab had a distinguished multiplicative effect, catalyzing regional PlanetLab-based initiatives worldwide, such as Global Environment for Network Innovations (GENI) in USA

---

[8] OSI stands for Open System Interconnection reference model [1].
[9] At the time of this writing, PlanetLab has 1080 nodes that span to 496 sites.

[68], OneLab2 in Europe, CoreLab in Japan and G-Lab in Germany. Also, VINI [69] extends PlanetLab to enable link layer VNs and maintains a "private" PlanetLab instance. In 2005, ORBIT radio grid testbed at Rutgers University [70] was deployed to address open wireless experimentation with scalability, controllability and reproducibility. More recently, "GENI initiative has further motivated efforts for ORBIT virtualization" [71] as well as started to prototype methods to integrate wireless testbeds with wired ones. See Chowdhury and Boutaba [61] for more details about virtual network testbeds.

Before we go any further, let's briefly look PlanetLab design principles [72]. PlanetLab aimed to provide large scale continuous experimentation based on real users traffic. Hence, it does not make sense carry out just one experiment at the time. Rather, PlanetLab used network virtualization to allocate a slice of the overlay network to each experiment. Each network node has a VMM that allocates and schedules processing, storage and transportation resources for each slice. Scalability is required to support a large number of simultaneous slices. Since the overlay network could be used by researchers and the general public, resource control should be distributed to deal with the conflict of interests between nodes owner and theirs users. Also, "overlay management services should be unbundled and run in their own slices, and APIs should be designed to promote application development" [72]. Peterson et al. argued that such Application Programming Interfaces (APIs) should become available to promote application development even at testbed. They contend that applications development is in general neglected in traditional testbeds. This is another reason why virtual testbeds should allow continuous experimentation, since users need time to try on new applications.

Aligned to PlanetLab, GENI project arose from the frustration of a group of researchers while trying to query against the infrastructure of the existing network and the difficulty of keeping the networks running properly [73]. These difficulties did not allow them to create new solutions (radical) to resolve existing problems, or stop the network in use to test new protocols. After several workshops on how to build a new infrastructure capable of supporting the development of solutions to old problems, GENI was launched in 2006 with NSF funding. The project aims to meet this need by building a large-scale infrastructure, able to allow testing of new protocols and stress them in scale with real traffic, end to end, making resources available to a large number of experiments in parallel through network virtualization. GENI project allows to test new architectures, which may or may not be compatible with the current Internet. Network virtualization allows to allocate resources to each experiment without affecting other ones. Each research can have its own slice of the testbed.

Going back to 2005, Peterson et al. revisited the original virtual testbed idea with a new focus to network virtualization: to use network virtualization to overcome the current Internet impasse. "... the status quo is no acceptable. We (as a community) are unable to deploy, or even evaluate, new architectures". The idea was to support several concurrent overlays to encourage innovative architecture changes, instead of providing ad hoc improvements. The authors argued that network virtualization can be a key aspect of the architecture itself, rather than just a way to evaluate new architectures. Thus, innovative architectures could gradually migrate from virtual testbed experimentation to commercial use. However, some coordination is required to allow interoperability and mutual benefit among overlays. This new approach not only encouraged research community, but also put network virtualization in the spotlight. Quickly, network virtualization is assuming a leading role in future Internet proposals, without forgetting their fundamental role in experimentation. Virtually, all current efforts on new Internet architectures consider virtualization as the tool to provide required generality, plethora, isolation, transparency and programmability of substrate resources. Examples are Akari [2], 4WARD[10] [74], FIA cluster Management and Service-aware Networking Architectures (MANA) [16] and USA Future Internet Design (FIND) [5][6]. Moreover, many manufacturers began to use virtualization to enable network programmability and customization [63][75], e.g. Cisco, HP, IBM, Juniper.

A virtual network consists of virtual nodes connected by virtual links, thus forming a virtual topology [61]. This means that not only nodes could be virtualized, but also links. Link virtualization enables to share one physical link among several virtual ones. Interestingly, Virtual Topology Design (VTD) is a well studied subject in the scope of WDM networks. In 1998, Ramaswami and Sivarajan explored VTD in their classical book about optical networks [76]. VTD and Routing and Wavelength Assignment (RWA) are notoriously complex problems (NP-hard). Haider et al. [75] argue "that the interactions between VNs and substrate network are far more complicated than the case of traffic flows and conventional networks". However, resource allocation is not the only one facet of VNs life cycle. VN life cycle looks more or less like the existing Session Initiation Protocol (SIP) session admittance in convergent networks, such as IP Multimedia Subsystem (IMS) [77] or Next Generation Network - Global Standards Initiative (NGN-GSI) [78]. First of all, it can be necessary a negotiation between virtual and substrate networks, in order to determine requirements, fairness, quality levels and network descriptors. Then, alike convergent networks traffic flows, a VN admission phase could be performed. At this phase, resources description and discovery will be necessary. SN available resources can be announced to facilitate admission control. If there are sufficient SN resources, a new VN could be instantiated, and resource reservation should occur. It is convenient to note again that as VTD and RWA problems in optical networks, VNs instantiation and resource reservation is a NP-hard problem too [79][61][75].

Virtual networks policing could be necessary to avoid misbehavior and unfairness. Scheduling and management of processing, storage and transportation resources should be done according to established contracts. Here, it is a very influential balance among utilization, isolation, fairness and quality. Similar to buffer management in converged networks, total isolation can lead to low utilization and high quality, while total share can lead to high utilization at the cost of quality loss and even unfairness. Thus, at least two approaches are possible: static or dynamic allocations [75]. In addition, VNs management and monitoring are also required. VNs life cycle should also include mechanisms for VNs modification, optimization and removal. Modification could be intentional or accidental,

---

[10] 4WARD is a project in the European 7th Framework Program (FP7) Call 1 and partly funded by EU.

whereas the former could be used to transparently overcome failure situations. Network virtualization could be a useful tool to provide network reliability [63], since SN failures could be overcome by changes in the virtual topology. Besides these technical aspects, there is also the dependence with possible new Internet business roles, which could affect several life cycle phases, especially in multidomain multioperator environments.

Another important requirement in network virtualization scenario is scalability. According to Carapinha and Jiménez [63], "the number of VNs on a specific network administrative domain may grow to the order of thousands, or more". Niebert et al. [80] argue that the ability to create VNs in scale is crucial to the success of network virtualization as a tool for network generality. Scalability of mechanisms for resources description, requesting, discovery, provisioning and management are serious concerns [80]. Resource description needs to be comprehensive, detailed and perhaps standardized. Resources request also needs to be homogenized. Resource discovery mechanisms could investigate available resources in a distributed manner. VN management requires new mechanisms alternatives, since many actions will be in the software environment. In a multidomain multioperator scenario, the security, privacy, and trust among parties have several challenges for network designers, some leaving technical realm. Like other architecture entities, VNs should also have unique identifiers [63]. Finally, there is the interoperability among VNs, which can compromise performance and increase complexity [80]. Since every VN could use different protocols, formats, etc, interoperability among them becomes challenging for multidomain multioperator SNs. A solution could be the adoption of design patterns, where some common "invariant aspects" are chosen by mutual consensus [80]. The requirements and mechanisms listed up to now, undoubtedly, ask for more autonomous operation in order to reduce OPEX and to improve quality, fairness, utilization, etc.

2.3.2 Wireless Network Virtualization

Nevertheless, substrate resources are not exclusive of wired links. The importance of wireless access, as well as RWI, IoT, CRNs, WSNs, etc, require that any proposal for substrate architectural generality should include the wireless environment. However, wireless network virtualization not only imposes more challenges for designers, but also turns VN support more complex [81]. The first challenge is to provide adequate coherence and isolation among VNs slices [82]. According to Smith et al. [82], coherence means that every transmitter and receiver of the SN should operate accordingly to an established channel access control. Isolation means that transmissions should be carefully scheduled to avoid collisions and interference among VNs. The second challenge is related to the physical uniqueness of wireless nodes, where heterogeneity and devices asymmetries are common. In 2006, GENI [83] provided a survey on possible channel access schemes for wireless virtualization: Frequency Division Multiple Access (FDMA) [1], Time Division Multiple Access (TDMA) [1], combined FDMA+TDMA, Frequency Hopping and Code Division Multiple Access (CDMA) [1]. Smith et al. [82], reexamined such approaches and proposed a TDM based wireless network virtualization scenario. Shrestha et al. [79], provided an approach for the creation of wireless slices in mesh networks using either Spatial-Division Multiple Access (SDMA), TDMA, FDMA or a combination of them. Stanford University OpenRoads project provided a different approach where commodity radio is virtualized into slices using flow-tables (similar to Multiprotocol Label Switching – MPLS) [84]. The third challenge regarding wireless virtualization is to support generalized mobility in VNs, dealing with users, terminals, nodes and other networks resources mobility. According to Söllner et al., this is one of the "major research topics in 4WARD" [81]. The mobility of a physical node has a direct impact on the VNs that pass by it. Mobility of virtual nodes impacts SN current processing, transport and storage allocations.

2.3.3 Virtualization Weaknesses

So far we discussed strengths of network virtualization. However, is there any negative point? The abstraction layer created by virtualization software can impact performance. Depending on the type of virtualization used the overhead of the indirection layer can be considerable. Take, for example, full virtualization, where virtual machines run over a VMM and a host OS. According to Marinescu and Kröger [60], such approach performance "can be up to 30% less than when running directly on hardware". Peterson et al. also complain about PlanetLab performance [62]: "It is clearly not possible for Planetlab nodes to compete with custom hardware. Similarly, the overlay's virtual links cannot compete with dedicated links". It is evident that to achieve superior performance it is necessary to keep virtualization overhead at a minimum. Cross-ETP[11] vision document [15] argues that "there is no definitive answer concerning negative impact that would result from the introduction of this new level of indirection".

Another point is security, privacy and trust. Since SNs most likely belong to different operators than VN ones, appropriate mechanisms will be needed to avoid security breaches and attacks. Another noteworthy point is OS and virtualization layer software security. An attack to any substrate hardware OS, hypervisor or any virtualization tool could be disastrous, affecting tens of VNs. Such problem has been relatively overlooked by current research. It is already well known that network security has in OSs a crucial gap. Finally, we must seize the opportunity that virtualization offers to build more efficient VNs in terms of control of information flows, avoiding the functional replication that exist today, taking advantage of cross-layer designs. However, the lack of mechanisms or common elements in SNs and VNs design can lead to interoperability and efficiency problems, similar to the ones we usually experience today with low compatible networks. In fact, according to Cross-ETP [15], "there is no proof so far that virtualization (that relies on the indirection principle) is resolving any of the FI technological challenges". More research is required.

---

[11] "Industrialists and academics, involved in several European Technology Platforms (ETPs), have come together to devise a strategy and action plan that will make the Future Internet an industrial, economic and societal success for Europe." [15].

3 Identification, Mobility and Localization

Another limitation in the current Internet is that IP addresses are used not only to identify nodes, but also to locate them in the network. It is necessary to separate identifiers (IDs) from locators to solve this problem. The identifier will be used to identify uniquely a network element. The locator will be used to locate this element in the network through the point it is attached. However, not only physical network elements must be identified, but also logical entities. Since the diversity of entities in a global information infrastructure is enormous (users, groups, services, applications, functions, storage, processing, machines, and interfaces, just to cite some), future networks must deal with identification and localization methods more holistically, in order to provide entities individuality in the network. Entities could be personalized, searched, localized, moved, modified, etc, while keeping their identity immutable. In such way, the network becomes fully transparent, since every entity has a unique identifier. This approach could improve localization, mobility, security, privacy, trust and accountability, as we will discuss in the next paragraphs.

Today, IP addresses are dynamically attributed to users' terminals to accomplish with the lack of unique IP addresses in the network. Therefore, IP addresses change frequently, mainly after a terminal reboots or moves in the network. The identity is preserved by means of user profiles in the operator's database. However, such frequent changes create difficulties for mobility and localization management in current networks due to inconsistencies, different domains, etc. When identifiers are decoupled from locators, it is possible to move things without "loss of identity". Thus, when a terminal moves from a geographic region A to B, locators change, but identifiers remain the same, allowing all the other functions to work properly.

In this scenario, a locator points out to an identity. Locators could be physical or logical. Physical locators allow determining the geographical location of a person, a machine or any other entity. Logical locators allow finding out where entities are in the context of a logical architecture. Therefore, some logical hierarchy must exist in the architecture, to relate logical locations with identities. Localization and movement of network entities, users, or machines, could be handled by localization mechanisms according to security, privacy, trust and accountability/anonymity[12] established policies [85]. This is necessary since localization and movement are sensible information. Today, IP's lack of transparency and IP semantic overload makes such mechanisms very limited.

Future networks must support general mobility as proposed by actual IP convergent networks, which natively support of all types of mobility: user mobility, terminal mobility, service and application mobility, functions mobility, or in general entities mobility. This explains why individual identifiers are required to a large set of entities: it is because we want to move, identify and locate them during

---

[12] Anonymity refers to the "absence of identifying information associated with an interaction" [86], while accountability refers to the ability to attribute someone's actions/information to an identity. It means to identify the source of information.

or after movement. Such ID/locator decoupling improves security, since identities remains the same independently from location. It also allows multihoming, which is the capacity of an identity to have multiple networks interfaces in order to increase availability and robustness.

However, the indirection step created by ID/location split brings not only benefits, but also notable challenges for security. Since security is required for all network entities, people and machines, unique identification allows to determine precisely the identity of information sources (a.k.a. information accountability) as well as the location and movement of such information sources. Such feature requires further multidisciplinary investigation, since it must be supported by national and international legislation, regulatory agencies and operators. However, information accountability could help to react to distributed security attacks where exploits are overspread in multiple operators, regions or even countries.

Identifiers and credentials[13] could be used together with authentication mechanisms to authenticate and authorize entities. Secrecy mechanisms, such as public cryptography, could also use unique identifiers to generate public keys and digital certificates. In the current Internet, security associations are established based on IP addresses. Whenever a terminal moves and its IP changes the security association related with this terminal could be compromised. Future networks could establish granted trust relations based on identities, thus improving security.

At this time, the reader could have already imagined the massive scalability problem this section' requirements bring to future networks scalability since the number of entities is immense and to identify them uniquely will demand innovative ideas. Therefore, there is a delicate drawback among identification, mobility and scalability. This is a tremendous challenge for future network designers. In summary, future architecture requirements include:

1. The need for unique identification of entities in order to improve security, accountability, mobility and localization.
2. The need for identity management to allow adequate privacy management.
3. The need for generalized mobility to improve users experience, diversity and flexibility of applications and services.
4. The need for a logical hierarchy in the architecture to relate logical entities identity with logical locations.
5. The need for scalability to support adequately identification, accountability and mobility requirements.
6. The need for new methods to generate identities and mechanisms to map such identities to locations and vice versa.
7. The need for authorized accountability in order to determine easily identity, location and trustworthiness of sources and sinks of information.
8. The need for credentials management and identity discovery mechanisms.
9. To investigate how identities could be used to support achievement of other requirements.

---

[13] "A credential is an attestation of qualification, competence, or authority issued to an individual by a third-party with a relevant de jure or de facto authority or assumed competence to do so" [87].

4 Adaptability, Autonomicity, Self-*, *-Aware and Manageability

In 2001, IBM worried about the increasing growth of their systems scale and complexity and has launched a manifesto on the impact of these limitations in the IT industry: "The company cited applications and environments that weigh in at tens of millions of lines of code and require skilled IT professionals to install, configure, tune, and maintain. Computing systems' complexity appears to be approaching the limits of human capability" claimed IBM researchers [88]. As time passed, these difficulties proven to be even worst when it is necessary to integrate solutions and make them available over the Internet. A proposed solution to the problem, as occurred in many other occasions in humanity, came from biology inspiration and was called autonomic computing. IBM researches realized that human autonomic nervous systems govern some vital background functions, such like digestion, heart rate, without the need to involve conscious processes in these tasks. Autonomic computing determines that IT systems must be able to govern themselves and meet high-level objectives proposed by the system operators. As described by Kephart and Chess in 2003 [88], the idea is to minimize human involvement by creating systems able to self-manage, i.e. with autonomic properties, capable to take care autonomously of their tasks, therefore, handling complexity and reducing life cycle costs. An autonomic system has four autonomic properties, which generically became known as Self-* properties [88][89][60]:

1. Self-Configuration – To configure automatically and seamlessly components and the system itself to achieve high-level goals. Manual configuration, installation, patching, upgrading will be no longer administrator tasks.
2. Self-Optimization – To optimize continuously and proactively system resources consumption and other aspects in order to improve performance, efficiency, quality, etc. Runtime parameter tuning will be autonomously made.
3. Self-Healing – To detect, diagnose and repair automatically localized problems and failures.
4. Self-Protection – To defend automatically against attackers, threads or cascade failures. The system can predict proactively potential problems based on collected data, such as logs, reports, etc.

All these properties are grouped under the term self-management, which is considered the essence of the autonomic computing [60]. Kephart and Chess [88] proposed that a self-management system would arise through the interaction among autonomous elements as well as with human operators. The autonomous operation arises as "social behavior" of the individual autonomous elements. Obviously, such approach is decentralized, distributed and cooperative. The autonomous elements are composed by two kinds of components: (1) one or more managed elements, which contains processing, storage and transportation resources to be automatically managed; and (2) an autonomic manager, which implements autonomic behavior inside the element. To perform autonomous operation inside autonomic managers, IBM researchers proposed a control loop similar to the already existing ones in traditional control systems. Such control loop was named Monitor-Analyze-Plan-Execute-Knowledge (MAPE-K). Monitoring of the managed element is done by "sensors, often called probes or gauges" [89]. Effectors are proposed to enforce autonomous manager decisions over managed elements. The collected data are analyzed and used to plan further actions. The objective is to accomplish human operator goals. Therefore, the focus of system administrators changes from a complete system operation to the establishment of high-level goals and verification of achieved results [60]. Internal knowledgement is used to achieve satisfactory autonomic operation. Transport resources are required to allow autonomous elements information exchanging.

According to Dobson et al. [90], the lack of consideration of autonomous elements communication as part of the problem is one of the most notable omissions from Kephart's and Chess's original vision. Interestingly, yet in 2003 David Clark and his colleges addressed the need for a new network research objective towards more autonomy in communication networks: "to build a different sort of network that can assemble itself given high-level instructions, reassemble itself as requirements change, automatically discover when something goes wrong, and automatically fix a detected problem or explain why it cannot do so" [91]. This proposal became known as knowledge plane and according to Dobson et al. [90], influenced Mikhail Smirnov to propose the notion of autonomic communications in 2004 [92]. According to Zseby et al. [93], Fraunhofer FOKUS institute established at this year a vanguard research initiative in autonomic communications focused in developing self-* properties for future networks. Autonomic communications refer to autonomous networks, capable to self-manage, to self-configure and to self-regulate [94]. It is a tendency to reinvent communication networks to cope with the increasing complexity, scale, diversity and dynamics of emerging architectures. The final vision of autonomic communications is to create entirely autonomous and adaptive devices, networks and services [94]. Although the advent of autonomic communication is near the appearance of autonomic computing, in literature, some differences in scope have been raised in [94][95]. However, more noteworthy is the assumption that in the context of new architectures, computing and communication resources should be treated holistically, i.e. does not make sense treat such efforts alone. Such jointly effort would be named autonomic information and communication technologies or autonomic ICT. In addition, the use of self-* properties only to implement self-management capabilities is a narrow view of the potential of this proposal. Self-control is an excellent example of extending autonomic ICT effort to include other portions beyond resource management.

In fact, self-* properties are already being considered to design not only built-in management, but also control, service enablement and orchestration in future Internet approaches. MANA cluster [16] defends the usage of self-* functions to allow a broad range of functionalities in the architecture. Self-* is envisioned to provide built-in FCAPS[14], enable design of self-configuring mobility frameworks and provide fault diagnosis and autonomous repair based on incomplete data. In addition, they propose autonomic management of global virtualized resources, self-adaptation of in-network management functions, self-contextualization and

---

[14] FACPS is the acronym for Fault, Configuration, Accounting, Performance and Security.

context awareness of resources, networks and services [16]. MANA work apparently has some influence from Autonomic Internet (AutoI) project also from EU. AutoI rests solidly in autonomic management to conceive new in-network management architecture for Future Internet (FI). The idea is to self-manage virtualized resources in order to achieve mobility, security, quality and reliability [95]. AutoI is composed by five distributed systems (or planes), namely: Orchestration, Service Enablers, Knowledge, Management and Virtualization (OSKMV). These systems collect operational information and promote *-aware[15] actions based on rules. AutoI management plane is composed by Autonomic Management Systems (AMSs) that run on top of VNs or other systems. It provides in-network management, *-aware mechanisms to monitor and determine network situation as well as achieved alignment with planned goals. Self-* capabilities are planned to provide FCAPS and self-control. In addition, it is responsible to manage, position and migrate virtual nodes.

The Akari project [2] also puts self-* properties in a prominent position at its architecture. Akari´s researchers argue that a network needs to be adaptive to evolve in a sustainable manner: "the network must be designed so that individual entities within the network operate in a self-distributed manner and that intended controls are implemented overall. In other words, a self-organizing network must be designed" [2]. Scale-free self-organizing control could be obtained by self-emergent autonomous actions at each node. In other words, network control emerges as a result of self-distributed local actions and communications, aligned to a large goal. EU EFIPSANS[16] project [96] is another FI initiative that envisioned self-* properties to automate FCAPS (and other control functions, such as routing, forwarding, etc). In this proposal, autonomic managers cooperate each other to achieve self-* features. EFIPSANS also claims for a holistic standardized reference model to help design and improving interoperability of autonomic ICT. The EFIPSANS Generic Autonomic Network Architecture (GANA) deals with complexity and effectiveness of control loops, their hierarchy, decision-making processes, objectives, goals, conflicts, policies, interfaces for human operators, multiple administrative domains, etc.

Another project that considered self-management from the design stage is 4WARD approach [97]. The in-network management is supported by composeability of autonomic entities, self-description of autonomic elements to achieve collaboration, self-security, self-accounting, interoperability while maintaining self-* capabilities, unique identification of autonomic elements to provide trustable operation.

Besides these projects, there is another project aimed to empower new wireless networks with a collaborative approach of self-management and cognitive radio networks. In January 2008, the End-to-End Efficiency (E3) FP7 project started to

---

[15] *-aware is a generalization of contextualized actions in ICT. For example, a service enablement platform is said network aware if it considers network condition on its actions. *-aware capabilities could be seen as cross-layer contextualization.

[16] EFIPSANS stands for Exposing the Features in IP version Six protocols that can be exploited/extended for the purposes of designing/building autonomic Networks and Services.

implement both cognitive radio networks and self-* capabilities, focusing on integrated, scalable and efficient multidomain, multioperator wireless networks [98]. Collaborative and autonomic elements behavior are combined to provide radio resource management, spectrum management and radio network self-organization and self-optimization.

Recently, Dobson et al. [90] revisited autonomic computing vision and argued that to achieve self-management "the system must be aware of its internal state (self-awareness) and current external operating conditions (self-situation), detect changing circumstances (self-monitoring), and accordingly adapt (self-adjustment)". This quotation shows the appearance of new self-* properties, which together with the so-called situation awareness[17] are in the spotlight lately. In 2009, Smirnov et al. [99] multi-author paper focused in demystifying self-awareness. The authors claimed that "self-awareness of autonomic systems is the only challenge that helps to rigorously and systematically address" the tangled hierarchy behind autonomic ICT. The authors determined aspects needing further investigation in self-awareness: control loops hierarchy, information contextualization, timely system adaptation, discovering of relevant contexts, behavior compose-ability and process assessment.

The first feature concerns to control loops diversity and dependency due to support to self-emergent composable functioning. A definition of self-emergent behavior is given by European Commission Report on Future and Emerging Technologies (FET) consultations [100]: "The concept of decentralised heterogeneous self-organised systems depends on specifying desired properties and behavior unambiguously so that the system can be persuaded to exhibit emergent behavior to satisfy requirements without a top-down or centralised control mechanism". Information contextualization is relevant to achieve situation awareness as well as to allow sound decisions. Relevant information needs to arrive timely in autonomic elements actuators. Self-contextualization is also required to determine in which context a piece of information is relevant. Self-emergent behavior could be created from other basic earlier behaviors. Finally, process assessment is required to gauge correctness, quality and fitness of decisions made as well as to feedback autonomous operations.

Self-awareness is considered to be specific to some ICT features. Smirnov et al. [99] propose the notion of a Self-Awareness Function (SAF), which enables to describe further relevant contexts to achieve self-awareness and situation awareness in both computing and communication aspects. The work pointed that autonomic elements need to know not only their internal state, but also their environmental condition. Network nodes collaborate with each other to determine network situation, therefore, achieving an increased level of situation awareness, i.e. an increased knowledge about network operation and environment condition [16]. The cooperation among network nodes is being pointed to address new architecture challenges as well as to achieve common goals and self-management property [93][89].

---

[17] There is still a lively debate about the true meaning of the self-awareness, situation awareness and self-situation.

MANA cluster supports the notion that self-awareness is necessary not only for self-management of computation resources, but also for communication ones [16]. MANA proposes an increased level of self-contextualization, self-awareness, self-composition, self-adaptation and context awareness to achieve effective self-emergent behavior. Cross-ETP vision document [15] argues that "if perfect situation awareness is achieved i.e. all the relevant factors for autonomic decisions are known and processed with respect to the decision-makers goal, then, the decision is evident". This statement advocates for the importance of network situation awareness. Therefore, everything indicates that situation awareness is a requirement to make consistent decisions towards self-management, and consequently towards successful network goals [93].

The problem of how to achieve self-awareness and situation awareness appears to be a key point in autonomic ICT. AutoI researchers [95] defend that one of the main research challenges behind autonomic ICT is to achieve self-awareness. A first glance at the problem reveals that detailed network information needs to be collected, filtered using self-contextualization mechanisms, distributed to the other relevant nodes in the right time. Thus, cooperation among elements appears to be fundamental to improve quality and scalability of information gathering. Another point is that the decision-making mechanisms often will have to work without perfect self-awareness and situation awareness. In addition, trust, security and privacy are other concerns, since information gathering needs consent [93] and could introduce opportunities to attackers. Therefore, trust relations would be established in order to allow collaboration among autonomic elements.

In summary, the scale and complexity behind new Internet architectures has surprised even the most optimistic. Just to name a few requirements that fundamentally contribute to this complexity consider generality, adaptability, diversity, security, privacy, trust, identification, location, mobility, management and deployment. Anyone who has read about or already encountered the problem of designing systems that can handle several of these requirements simultaneously often has the feeling of failure against the problem. So imagine the "headache" that will be to put these systems to operate together. It seems a mission impossible, with very high Capital Expenditure (CAPEX) and OPEX. To think only on people to integrate and operate these systems extremely complex, distributed, virtualized, in multidomain and multioperator environments, is to be on the border of viability. To maximize business potentials and society wishes while maintaining excellence in quality is a tremendous challenge for these systems. Therefore, it is clear that we need more autonomic operation than in actual solutions. Current Internet is almost entirely operated and managed manually. Convergent networks, like NGN-GSI [78] and IMS [77], are already putting more and more pressure on IT and Telecom departments of operators [90]. Hence, it makes sense think in a new Internet that advances the furthest in the direction of autonomy and adaptability, and perhaps even in the direction of cognitive ICT approaches. Examples are the human brain inspired ICT [100], biological inspired ICT [101], new results of artificial intelligence [102][103], cognitive informatics [104], cognitive computing [105] and other "consciousness" like approaches.

Future network architecture requirements regarding adaptability, autonomy, self-*, *-aware, controllability and manageability include:

- Innovative autonomic and cognitive approaches to minimize human intervention and design complexity as well as to increase scalability, controllability, manageability, security, and privacy.
- New approaches to assess realizable goals as well as coherent instructions to the network.
- Establishment of trustable collaboration and self-emergent autonomic elements behavior.
- Looking for increased levels of self-contextualization, self-awareness, self-composition, self-adaptation, context awareness and situation awareness to achieve successful distributed functionality.
- Designing of built-in control, management, service enablement, security, privacy, trust, contextualization and orchestration.
- Designing of dynamic and hierarchical control loops.
- Measurement of the quality and suitability of the decisions made to quantify obtained functioning.
- Adaptability to dynamic conditions in requirements, goals, information and decision processes.
- Integration of generality, adaptability, autonomicity and scalability in design.
- Advancing the furthest in the direction of autonomy, adaptability and generality and maybe even in the direction of cognitive ICT approaches.

5 Semantic, Context, Naming and Routing

In its early days, Internet was focused in creating end to end host connectivity via links and routers. As time passed, the interconnection network was losing attention to the endpoints and their applications, which eventually led to the emergence of the World Wide Web (WWW) [1] and the popularization of the Internet. Such movement has led the transformation of the Internet in the main vehicle for information exchange and is increasingly changing the way we produce and consume content [57]. Despite such success, many Internet researchers are now pointing towards an information-centric paradigm [106][107][108][109]. The main reason is that current endpoint-centrism led not only to notable success, but also to important limitations, mainly regarding information treatment in the network. For example, there is not a persistent information naming scheme to provide host decoupled information identification, naming and location [110]. What happens today is that content is identified and located based on Uniform Resource Locators (URLs) and IPs. Hence, the same content (e.g. a movie file) is identified and located by different URLs. It is a problem to move this content from one domain to another, which frequently requires additional control to deal with HTTP redirects [110]. The information dependence on host location hinders applications and services evolution [106]. Also, the network contains too few mechanisms to help content distribution, caching and transcoding. With few exceptions, all the work is left to endpoints/applications. There is no support for anycast communications, i.e. distribution of the nearest copy of a certain content. The information is not represented in a consistent manner in the network, e.g. content lacks on metadata to describe them adequately. It is difficult to manipulate information in a contextualized manner as well as to deal with identical copies. To enable customized experience, we need "to abstract from the syntax to semantics" [111]. In addition, the network lacks on securing information [106][110]. We are not sure whether we can rely on content received over the network. We have neither the right to deny the receipt of certain unwanted information. Also, there is no support to different communication models, such as delay tolerant, publish/subscribe, etc. In summary, these and other limitations have attracted increasingly interest in rethinking network architecture from an information point of view.

A new Internet architecture is a fantastic opportunity to consider information as a key ingredient in design, since everything can somehow be seen as information, e.g. identity, policy, contracts, etc [108]. It is also an opportunity to push Internet towards a semantic web as advocated by Sir Tim Berners-Lee in 2001 [112]. His idea is to define the meaning for information and its treatment, so that the web can "understand" what people and machines want. It means to build an autonomous knowledge web, including context-aware applications and services composition. As one would expect there is today a tremendous diversity of visions, proposals and designs aimed to create new approaches not only focused on information, but also in semantic, autonomy and service composition [113].

Many information-centric efforts have appeared since 2006 and nowadays there is a substantial agreement that information-centrism promises to solve some of the host-centric limitations in a consistent manner. Some efforts quite mentioned in the literature are EU FP7 projects 4WARD NetInf [109][107] and Publish Subscribe Internet Routing Paradigm (PSIRP) [108], as well as USA Content Centric Networking (CCN) [106] from Xerox Palo Alto Research Center. Some efforts from the semantic point of view are World Wide Web Consortium (W3C) Semantic Web [114] and EU FP7 projects Service Web 3.0 [115] and Service Oriented Architectures for All (SOA4All) [116]. W3C Semantic web defines a stack composed by several standards and tools, such Extensible Markup Language (XML), Resource Description Framework (RDF) and Web Ontology Language (OWL).

4WARD deliverable D6.2 [117] provides a comparison among information-centric proposals. Although there are some noteworthy differences between them, similarities also occur. NetInf, PSIRP and CCN proposals persistently represent information independent of copies, location and encoding. Achieved representation contains information-specific metadata, e.g. signature, semantic, access rights and other attributes [109]. Information representation is decoupled from networks/hosts, creating an indirection level between information and its treatment. In NetInf approach, such representation is called Information Object (IO). Information representation and identification has a strong impact on architecture scalability, since there is a tremendous quantity (exabytes) of information to be represented. Current efforts are adopting separate mechanisms to identify information, people and network entities. Apparently, it makes sense think in treating such identification problems in a more holistic way.

The objective of a name resolution scheme is to find out locators for a named content. The NetInf, PSIRP, and CCN use such schemes to put names on IOs. They also enable information location independently of where it is stored. Two kinds of name schemes are being considered: hierarchical and flat (non hierarchical). NetInf uses a structured flat naming scheme for IOs. Each IO has a global unique name or Identifier (ID) [107]. Opaque flat names are sometimes criticized because they are not easily used by people. To avoid that, CCN uses a hierarchical naming scheme composed by a global-routable name, an organizational name and a series of automatic version and segmentation components, creating a unique publisher rooted naming tree [106]. PSIRP uses two levels of identifiers [109]: rendezvous and forwarding. The Rendezvous iDentifier (RiD) is used to implement publish/subscribe paradigm, i.e. to put in contact information publisher and subscriber. Forwarding iDentifier (FiD) enables information forwarding after rendezvous [117]. Also, related to names, two issues that deserve attention are: network entities naming and alignment with semantic web proposal [112]. More investigation is required to answer questions like: what is the need of network entities naming, e.g. things, services, applications, etc. If network entities will be named, is it possible to design common mechanisms for naming purpose? What is the alignment degree to be achieved between information-centric and semantic web approaches? How search and discovery could take advantage of the relationships among naming, semantic, identity and location?

The publish/subscribe communication model [119] was adopted somehow in all previous cited proposals. In NetInf approach, IOs are published, and subscription is done by means of queries for name resolution system [109][107][117]. In CCN, name prefixes for named data are published and subscription is done by routing Interest packets to the publisher [117][106]. The rendezvous process put in contact publisher and subscriber in PSIRP approach. PSIRP allows subscribers to send Interest packets for data not yet published. The publisher can create data on demand to satisfy user needs [117][108]. Although publish/subscribe model is being adopted by all these proposals, new architectures need to maintain neutrality and transparency to allow innovation, i.e. this model should not be the only one possible.

Regarding security, all proposals take the opportunity to rethink content security from the information point of view. According to Smetters and Jacobson [120], "content-based, rather than connection-based, security would allow users to retrieve securely desired content by name, and authenticate the result regardless of where it comes from". NetInf provides owner authentication using digital signatures in metadata with the involvement of a public certification authority. NetInf IOs names can contain a cryptographic hash function of the data to provide self-certification [117]. Self-certifying names have the advantage to be generated without third-party involvement [120]. However, opaque names can require mapping mechanisms to facilitate usage by people [120]. Thus, an attacker can infiltrate false content at this point. To deal with this problem, CCN "authenticates the linkage between names – arbitrary names, including user-friendly ones – and content rather than authenticating the content or its publisher" [120]. In CCN, public-key signatures enable to authenticate name-content binds in such way that anyone can verify its authenticity [106]. PSIRP security differs for real-time or stored contents [58]. For stored content, PSIRP creates the rendezvous identifiers applying a hash calculation over the data content. For real-time content, public keys compose the rendezvous identifier [58]. PSIRP also uses packet level authentication. Because all information is associated to an originator, information security must be integrated somewhere in architecture with network entities security. Privacy and accountability are also related as will be discussed in Section 6.

According to Van Jacobson [106], "99% of the Internet traffic today consists of named chunks of data, e.g. video, P2P, web, etc". He argued "that named data is a better abstraction for today's communication problems than named hosts". This proposal has gained more and more supporters and transport based on named content is an emerging trend. The objective is to route/forward/cache information previously located by a naming resolution scheme. NetInf uses two name-based routing approaches (see [117]): Multiple DHTs (MDHT) and Late Locator Construction (LLC). MDHT enables routing of data requests by name and routing of desired information using shortest or reverse paths. MDHT works with hierarchical domains [109]. LLC uses "hierarchical locators constructed on demand" [109] to achieve an end to end routing based on ingress, core and egress domains. PSIRP routing is done on demand when rendezvous among publisher/subscribers occurs. The rendezvous system requests to a topology system the elaboration of a FiDs tree from publisher to interested subscribers. The topology system configures fast forwarding tables at network nodes to establish a direct path from publisher to subscribers. Data forwarding relays on FiDs and RiDs. In CCN, each desired content packet needs to be requested by an interest packet. After matching an interest name with a content name in every node, the content packet routes back to the subscriber using the reverse path. It is up to the subscriber resend interest packets in case of timeout.

In summary, main challenges in information-centric approaches are to provide efficient, scalable, fast, secure, mobile ready, and global named content localization, routing, forwarding, and caching. Network caching is a powerful mechanism to improve performance. However, it has impacts on privacy and legal aspects. For example, who is responsible for inappropriate contents cached on some network elements? What happens if some cached content is inadvertently sent to an unauthorized subscriber? Still, according to MIT Privacy & Security Working Group [120], "is the rendezvous valid? Is there a match between attributes and interests, and who is validating that?" These issues illustrate the need for further investigation on the relationships among information and networking security, privacy and trust. Semantic also poses critical challenges for new architecture designs, mainly in terms of scalability, ontology[18], information relevance and semantic generation [113][116][111].

6 Security, Privacy, Trust, Transparency, Anonymity, Accountability and Safety

There is a solid consensus that privacy and security have critical deficiencies in the current Internet. In the beginning, Internet was controlled by a small group of

---

[18] There exists many definitions for ontology. Gruber [121] defines an ontology as "a formal explicit specification of a shared conceptualization for a domain of interest".

institutions with limited access. Computers and routers are assumed to be trustable. Today, Internet is a worldwide open access network where trustable computing systems are the dominant minority. The range of privacy and security vulnerabilities and their exploits is so vast that is difficult even to enumerate: computer viruses, worms, trojans, spyware, dishonest adware, phishing, spam, spoofing, code injection, frauds, etc. Not only are end users or applications attacked, but also databases or even the Internet. Users are complete lost when an operational system or firewall software open a window to present some incomprehensive message regarding a possible threat or vulnerability. There is no base for decision! Intrusion and Deny-of-Service (DoS) attacks have already compromised availability in some telecommunications operators, generating loss of revenue, fines and breaking of Service Level Agreements (SLAs). The problem is becoming worst, since new technologies and applications, such as sensors networks, video applications, instant messaging, presence and location services, are increasing human behavior monitoring. They are additionally bringing sensible information to the network. Real and virtual worlds information are being massively captured (or even stolen) and used without authorization on the Internet, causing loss of privacy, freedom and other damages. Sooner or latter virtual attacks could produce real-world damage in our homes, cars, offices, etc, by means of hijacked actuator networks.

To guarantee privacy and security for users, applications, machines and other entities, while maintaining scalability, openness, diversity, heterogeneity, extendibility and flexibility is a tremendous challenge. The interrelation and interdependency among those requirements is so complex that diversity and opposition of visions, ideas and debates are quite notable. However, some common requirements appeared in current research regarding future Internet privacy and security improvement. First of all, it is desirable that security, privacy, trust and accountability requirements must be built-in (or inherent) features of future network architectures. It means that such aspects must be considered from the beginning, to benefit all the architecture. Also, there is consensus that people must trust not only in the network, but also in its entities.

In the current Internet, too much information is send to the destiny users without previous authorization, e.g. email spam, pop-ups, malicious cookies, etc. In future networks, other communications models could be explained [122]. In consented communications, information is exchanged only if receiver authorizes. In this case, trust relations must be established seamlessly among users (or entities acting on behalf of users). Consented communications between trusted entities, services, and users could provide improved levels of security and privacy. Trust among autonomous elements is also a concern. The dependability on different trusted parties could be determined, and intuitive risk announcements could be propagated for users as well as for other entities according to appropriate contexts. Trust negotiation mechanisms would be necessary to automate decisions, facilitating user's life and improving efficiency. Also, reputation monitoring mechanisms are necessary to determine everything degree's of confidence, from sensors to media servers. Identity, credentials, and trustworthiness must be established seamlessly in the network. In addition, usability and intuitiveness of trust methods and mechanisms must be considered. The precise relation between identities, credentials and reputation needs to be determined as well as their management. Reputation policies could be used to increase threat monitoring of malicious untrustworthy entities.

Another concerning feature is privacy. Cross-ETP vision document [15] defines privacy "as the right to informational self-determination, i.e. individuals must be able to determine for themselves when, how, to what extent and for what purpose information about them is communicated to others." Some mechanisms necessary to deal with this definition are: (i) how to help users to protect their Personally Identifiable Information (PII) intuitively; (ii) how to manage trust relations in order to improve privacy; (iii) how to manage multiple identities and credentials; (iv) how to support anonymous identities and others Privacy Enhancing Technologies (PETs). PII is all the information that can identify a person uniquely, e.g. a biometric record, names, digital identities, etc. PETs are tools that help users to protect their PII.

Accountability is present only to a certain degree on the current Internet. Today, IP addresses are dynamically attributed to user's terminal to accomplish with the lack of unique IP addresses in the network. The IP addresses change frequently, mainly because reboot or mobility in the network. Therefore, if a user sends malicious content and causes damage to someone else, the identification of this user requires access to the operator database in order to determine which user was using that IP address at the moment of the threat. In general, this process is slow and requires the participation of a regulatory agency as well as network operators, according to existent national cyberspace laws. The authors of [15] say that the number of users and computers in the current Internet "makes anonymity a powerful weapon for security breaks of any sort". Therefore, some level of monitoring could be necessary to determine clear malicious behavior or inadequate use. However, the architecture must provide the means for privacy and anonymity. Here, is another delicate point where the debate is still happening.

Observe that when anonymity is complete, accountability is impossible. The reason is that it is impossible to identify the information source. Contrarily, when accountability is complete, anonymity is impossible, since information is monitored and the source is always identifiable. Future networks must be open to the diversity and heterogeneity of applications and uses. This means that future networks must provide not only anonymity for users that require it by force of law, but also accountability, when it is legal. For instance, a presence service could generate some information about a certain user presence in a shopping center, and inadvertently makes it available to other unauthorized web service. In this case, the presence service is guilty of such privacy violation. Therefore, some autonomous privacy enhancement mechanism could detect such a situation. However, what could happen after this detection is a matter of substantial controversy, because case law on computers and Internet privacy is still evolving. Other accountability examples are to identify/locate sensors generating a fire alarm, a user that is calling for help in a catastrophe, and so on. In summary, accountability is a requirement in future networks as well as anonymity. The question is how to design the architecture open and flexible to deal with both contradictory requirements, while maintaining network evolvability.

Obviously, usual security mechanisms could still be applied for securing future networks. However, new solutions are required to deal with such complex set of requirements. Undoubtedly, more autonomy is necessary to create proactive self-securing, self-protecting, self-monitoring, self-healing mechanisms, in order to detect and react against vulnerabilities, threats, intrusions, privacy violations and distributed attacks. Self-protection mechanisms could autonomously configure protection mechanisms like firewalls, filters, etc. Self-monitoring tools could monitor threats and violations against authorization, authentication, secrecy, integrity, and trust mechanisms. They could determine if some illegitimacy is occurring, save proofs of malicious or inadequate behaviors, and start legal actions, if any. Self-healing mechanisms could react against attacks, vulnerabilities, breaches, violations, installing patches, modifying firewall configurations, etc.

To summarize, future network architecture requirements regarding security, trust, privacy, anonymity and accountability include:

- To deal with the tussle [123] among privacy, accountability and evolvability.
- The need for privacy mechanisms in order to establish consented communications.
- The need for authorization control to make information available just to authorized parties.
- The need to improve trust on Internet: people should trust on the Internet and trustable relations among entities must be established to avoid/detect threats, vulnerabilities and violations.
- Mutual authentication and trustable communication establishment.
- The need for robust and scalable security mechanisms to deal with distributed massive attacks, including detection and accountability when legally authorized.
- To incorporate as much as possible securing mechanisms (to provide confidentiality, integrity and peer authentication) natively at all network levels: functions, terminals, nodes, services, applications, contents and other entities.
- The need for legal authorized accountability in order to track actions performed by entities.
- The need for mechanisms related to information, privacy, trust and reputation life-cycles.
- More autonomy on network security mechanisms, i.e. to develop self-secure and self-protecting networks.
- Measurement, analysis and classification of risks, vulnerabilities, and threats.
- To improve security, trust and privacy mechanisms to deal with other architectural requirements.

7 Neutrality, Openness, Diversity, Extendibility, Flexibility and Usability

One of the central design principles of current Internet was the end-to-end principle [2][124]. It states that application level functionality can not be placed at the network layer. Therefore, it pushed all application functionality to the end hosts, leading to the already discussed host-centrism (see Section 5). As a consequence, IP design was kept to a minimal, creating what frequently is referred as a "dumb network" with "smart hosts" model. This principle introduced network applications neutrality and the Internet potential to innovation was maximized. Internet was kept open to allow application diversity, e.g. FTP, telnet, email, etc. As cited by Akari [2], the WWW is perhaps the most significant result of such approach. Importantly, this history could repeat itself again. Furthermore, the most important use cases, business processes, applications and services of the future Internet could be completely unknown right now. Therefore, future network architectures should be generically designed without assuming preferred scenarios, since they are so unpredictable at design moment that to consider current speculations could reflect in a limited approach. A generic (usage independent) information network is required.

In this scenario, a diversity of evolvable, extendible and flexible software frameworks could be designed to co-exist over such generic information network. The "doors" should be open to satisfy future society needs, allowing unthinkable, rich, interactive and immersive scenarios of use. This usage/information decoupling favors network evolvability, since we do not know when we will be able to replace the Internet again [2]. Such approach requires that substrate resources, such as transportation, processing, storage and others are somehow exposed to overlying infrastructures in order to bring to live services and applications as information treatment processes. Therefore, substrate network generality is required. As discussed on Section 2, network virtualization approach is the current trend for generality. Interestingly, virtualization allows customizing entire sets of virtualized resources accordingly to service needs. Additionally, virtualized resources could be grouped to set up service aware customized networks. It means that underlying network resources, such as forwarding, queuing, scheduling, could be customized to create customized virtual service networks.

State-of-art in software design and computing, such as cloud computing, high-performance computing, grid computing, service oriented computing [125], autonomic computing [88] are candidate technologies to build overlaying software frameworks. They need to be scalable, flexible, secure, trustable and extendible as well as they must provide high-performance applications execution [126]. Another requirement is that software depends on other software, e.g. operational systems, bios, microcode, compilers, linkers as well as on computing and storage resources. Hence, dependability is a problem and will require appropriate treatment.

Recently, software design suffered a paradigm change from component based to service oriented design [15][127]. Service oriented software is based on the design paradigm that software applications can be flexibly and dynamically constructed by the composition of other network distributed software services or utilities [127], forming what is being called a Service-Based Application (SAB) [126]. Figuratively speaking, the idea is to make a software pyramid, where the software on top of the pyramid depends on several other composed software until reaching the pyramid base, where lays some fundamental software services or utilities. Such idea is also called Service Oriented Computing (SOC) [125] or Service

Oriented Architecture (SOA) [128]. Hence, there is a certain consensus that a service-centric approach will be adopted to develop the top portion of the new Internet, creating large-scale multi-enterprise complex service networks [126]. A service-centric approach assumes that "above a certain level of abstraction everything can be viewed as a service leading to the concept of the Internet of Services" [129]. Substrate resources, such as transport, computing, storage, could be virtualized as fundamental services to service-oriented infrastructures and frameworks. Fundamental services similar to that we use today, such as search, localization, geo-information and social networking could become available to compose other services dynamically.

The service-centric approach has some important implications and requirements [126][127][125][130]. Service-based applications could be developed, deployed and discontinued by a dynamic service life-cycle. The life-cycle starts when a client requires the invocation of a new service-based application. The first step is the search for adequate services to compose the final application. This phase requires discovery and selection of federated services distributed over different domains, providers or other third parties. This means that a service-based application could be composed by third parties software, no longer under control of developers as in traditional component based or desktop based applications [127]. To achieve this approach, methods and mechanisms for the seamlessly service describing, publishing, discovering and negotiating are necessary. Service describing must contain important information about a service in order to facilitate its selection. According to the MANA cluster [16], examples of attributes are "capacity, throughput, QoS, latency, protocol support, availability, security, etc., in a consistent format. They need to express cost and availability, scalability, and potentially elasticity and support for usage variations". This service information must be published in divulgation services in order to allow adequate service discovery and selection. Once a suited service is found, negotiation of a Service Level Agreement (SLA) takes place between parties. In the context of the European Network of Excellence in Software Services and Systems (S-Cube), this step is called service binding [130]. From this point on, the process resembles the existing steps in traffic flows admission at IP converged networks. The SLA will impact third-party resources. Thus, an admission phase is important. If admitted, the bind will reflect on resources reservation at third-party software infrastructure. Then, service installation and configuration proceed. Notice that the third-party software could be already running in some operational system at this time and shared among other services. SLA policing is necessary to assure quality, as well as service monitoring, logging and exception handling. Service management in terms of several aspects is required, such as pricing, failures, availability and resiliency. Finally, when the application is turned off by users or machines, service finishing is required to free up resources and closes SLA. Additionally, service adaptation could be necessary to change application functioning as desired by clients as well as service migration in case of inadequacy, failures or usage variations.

Besides services life-cycles, a more holistic view could include application and business process life-cycles integrated to services one. This could allow more complete application development scenarios, where services are composed to form applications, and in turn applications are created to give life to business processes [126]. More investigation is required to determine the benefits of this integration into a new Internet and into its businesses models.

From user's point of view, new Internet overlaying software infrastructures could improve network usability, which is the ability to provide good Quality of Experience (QoE). They allow application personalization and contextualization, creating diverse, rich, interactive and immersible experiences. They could provide user's self-servicing capabilities in such way that user's can configure themselves exactly what they want, when, where and in which payment model. In addition, it is desirable that applications could vary their functioning depending on user preferences and context. A service-based application could modify underlying services parameters or even change composing services to better fits user requirements. Contextualization means that user/application interaction is context-aware. In other words, user/application interaction depends on user preferences, physical situation, social networking, relation to real world, history, skills, connectivity, and other information – that could be used to characterize the situation of an entity [130][15]. Context-awareness could also provide semantic invocation of applications and services, as mentioned in [15]: "services can be flexibly detected and invoked based on semantically rich inference rules relying on properties describing context".

Usability could also be improved by means of personalized, elegant and clear interfaces. Ultimately, users will be able to design their own services and applications and to export them to other users by means of services mash-ups or scripts [15]. With this solution, future Internet could allow new applications plug-and-play development and deployment. Possibly users will can easily, quickly and flexibly develop and deploy their own applications in a self-service manner. These perspectives underlie a certain consensus that diversity and heterogeneity of applications and services will be tremendous on the FI. Millions of new applications could become available, compete and evolve together.

To support such scenario new Internet architectures could address:

- To keep open the possibilities for future use cases, applications and services, improving network extendibility and evolvability.
- To support diversity of use cases, applications, services, protocols, functions, etc, improving network flexibility.
- To allow new applications/services become easily, quickly and securely available, facilitating Internet extendibility and flexibility.
- The need to provide open, extendible and flexible support to a huge number of concurrent services/applications without compromise scalability, security, privacy and capacity.
- The need for innovative frameworks to orchestrate services/applications based on resources as well as to support seamless end-to-end private, trusted and secure communications.
- To make applications and services consistent and available anywhere, anytime.

8 Simplicity, Sustainability and Evolvability

The famous KISS (Keep It Simple Stupid) principle [124] is frequently cited as one of the principles which guided current Internet development. It states that design must be kept simple and unnecessary complexity should be removed. Apparently, design simplicity is not being pointed as a fundamental requirement in future Internet efforts as one could expect. Several proposals defend simplicity in architecture, but only a few are committed to keeping the design simple from the beginning. Akari [2] is one of such proposals. It argues that a network must be designed to be simple and for this reason it adopts the same KISS principle. Akari also aims to design a sustainable network, capable to support information society needs in the next decades. Simplification of integrated technologies is one of the most important Akari design principles.

Apparently, simplicity is achieved by refinement/evolution. To illustrate how difficult it is to design with simplicity one can evoke Leonardo Da Vinci's impressive quote: "simplicity is the ultimate sophistication". Future networks' architectures are not an exception. Therefore, it is desired that future network architectures could provide mechanisms for simplicity and sustainable evolvability, i.e. to provide means to allow new approaches to replace established ones, reducing unnecessary complexity and increasing efficiency and quality. From the sustainability point of view, to try to anticipate deterministic solutions for all possible tussles [123] in a future Internet is an ungrateful task, since the probability to miss the point is huge. A better approach could be to create a "digital savannah" where the evolution could take place, instead of trying to anticipate all the conflicting aspects at the design phase [131]. Since future networks have a complex, interdependent, multidimensional and multidisciplinary set of requirements, perhaps a Darwinian architecture approach [132], where network evolution can occur, would lead to better evolvability, sustainability and simplicity.

Acknowledgments. I would like to acknowledge Tania Regina Tronco for the invitation and opportunity to write this text. I would also like to thank Tania and Christian Esteve Rothenberg for the valuable feedbacks. Finally, I would like to thank Instituto Nacional de Telecomunicações (INATEL) for the support.

References

1. Tanenbaum, A.S.: Computer Networks. Prentice-Hall, Englewood Cliffs (2002)
2. AKARI, New Generation Network Architecture AKARI Conceptual Design. Project Description v1.1 (2008)
3. Pettit, J.L., Casado, J., Lockwood, M., McKeown, N.: Prototyping Fast, Simple, Secure Switches for Ethane. In: 15th Annual IEEE Symposium on High-Performance Interconnects, Stanford, USA
4. http://cleanslate.stanford.edu/index.php
5. Fisher, D.: US National Science Foundation and the Future Internet Design. ACM SIGCOMM Computer Comm. Review 37(3), 85–87 (2007)
6. http://www.nets-find.net

7. Roberts, J.: The Clean-Slate Approach to Future Internet Design: A Survey of Research Initiatives. Ann. Telecommunication 64, 271–276 (2009)
8. http://www.usatoday.com/tech/columnist/kevinmaney/2005-07-05-famous-quotes_x.htm
9. Saracco, R.: Telecommunications Evolution: The Fabric of Ecosystems. Revista Telecomunicações INATEL 12(2), 36–45 (2009)
10. Jia, D., Duan, Y., Liu, J.: Emerging Technologies to Power Next Generation Mobile Electronic Devices Using Solar Energy. Frontiers of Energy and Power Engineering in China 3(3), 262–288 (2009)
11. Bullis, K.: Graphene Transistors. MIT Technology Review, Cambridge (2008)
12. Clemens, W., Fix, W., Ficker, J., Knobloch, A., Ullmann, A.: From Polymer Transistors Toward Printed Electronic. Journal of Materials Research 19, 1963–1973 (2004)
13. NTT, Special Feature: Silicon Photonic Technologies Leading the Way to a New Generation of Telecommunications. NTT Technical Review (2010)
14. EIFFEL Think-Tank Consortium, The Future Networked Society. Evolved Internet Future for European Leadership (EIFFEL) White Paper (2006)
15. Cross-ETP, The Cross-ETP Vision Document. European Technology Platforms (ETPs) Cross Vision Document v1.0 (2009)
16. Galis, A., Abramowicz, H., Brunner, M., et al.: Management and Service-aware Networking Architectures (MANA) for Future Internet: System Functions, Capabilities and Requirements. MANA Position Paper v6.0 (2009)
17. Presser, M., Daras, P., Baker, M., Karnouskos, S., Gluhak, A., Krco, S., Diaz, C., Verbauwhede, I., Naqvi, S., Alvarez, F., Fernandez-Cuesta, A.: Real World Internet Position Paper (2008)
18. Weiser, M.: Hot Topics: Ubiquitous Computing. IEEE Computer 26(10), 71–72 (1993)
19. Gluhak, A., Bauer, M., Montagut, F., Stirbu, V., Johansson, M., Vercher, J., Presser, M.: Towards an Architecture for a Real World Internet. Towards the Future Internet. IOS Press, Amsterdam (2009)
20. Meng, X., Xu, Z., Zhang, B., Huston, G., Lu, S., Zhang, L.: IPv4 Address Allocation and the BGP Routing Table Evolution. ACM SIGCOMM Computer Communication Review 35, 71–80 (2004)
21. Naoya, W.: Research and Development of 160 Gbit/s/port Optical Packet Switch Prototype and Related Technologies. Journal of the National Institute of Information and Communications Technology (NICT) 53(2) (2006)
22. Nakazawa, M., Yamamoto, T., Tamura, K.: Ultrahigh-Speed OTDM Transmission Beyond 1 Tera Bit-Per-Second Using a Femtosecond Pulse Train. IEICE Trans. on Electronics 85(1), 117–125 (2002)
23. Knight, J.C., Birks, T., Russell, P., Atkin, D.: All-silica Single-Mode Optical Fiber with Photonic Crystal Cladding. Optics Letters 21(19), 1547–1549 (1996)
24. Yeh, P., Yariv, A., Marom, E.: Theory of Bragg Fiber. J. Opt. Soc. Am. 68, 1196–1201 (1978)
25. Ranka, J.K., Windeler, R.S., Stentz, A.J.: Visible Continuum Generation in Air–Silica Microstructure Optical Fibers with Anomalous Dispersion at 800 nm. Optics Letters 25(1), 25–27 (2000)
26. Wadsworth, W., Ortigosa-Blanch, A., Knight, J., Birks, T., Martin Man, T., Russell, P.: Supercontinuum Generation in Photonic Crystal Fibers and Optical Fiber Tapers: a Novel Light Source. JOSA B 19(9), 2148–2155 (2002)
27. Hall, J.L.: Defining and Measuring Optical Frequencies. Nobel lecture (2005)

28. Udem, T., Holzwarth, R., Hänsch, T.W.: Optical Frequency Metrology. Nature 416, 233–237 (2002)
29. Miyagawa, Y., Yamamoto, T., Masuda, H., Takara, H.: Over-10000-channel 2.5 GHz-spaced Ultra-Dense WDM Light Source. Electronic Letters 42, 655–657 (2006)
30. Sergienko, A.: Quantum Communications and Cryptography. CRC Press, Boca Raton (2006)
31. Elliott, C., Colvin, A., Pearson, D., Pikalo, O., Schlafer, J., Yeh, H.: Current Status of the DARPA Quantum Network Report (2003)
32. Schmitt-Manderbach, T., Weier, H., Fürst, M., Ursin, R., Tiefenbacher, F., Scheidl, T., Perdigues, J., Sodnik, Z., Rarity, J.G., Zeilinger, A., Weinfurter, H.: Experimental Demonstration of Free-Space Decoy-State Quantum Key Distribution over 144 km. Phys. Rev. Lett. 98, 010504 (2007)
33. Masahide, S.: Overview of Quantum Info-Communications and Research Activities in NICT. Journal of the NICT 53(3) (2006)
34. Westmoreland, M., Schumacher, B.: Quantum Entanglement and the Non Existence of Superluminal Signal. Los Alamos National Laboratory Report (1998)
35. Ursin, R., Jennewein, T., Aspelmeyer, M., Kaltenbaek, R., Lindenthal, M., Walther, P., Zeilinger, A.: Quantum Teleportation across the Danube. Nature 430, 849 (2004)
36. Schmid, C., Kiesel, N., Weber, U., Ursin, R., Zeilinger, A., Weinfurter, H.: Quantum Teleportation and Entanglement Swapping with Linear Optics Logic Gates. New Journal of Physics 11, 033008 (2009)
37. Hourcade, J.C., Saracco, R., Wahlster, I.N., Posch, R.: Future Internet 2020: Visions of an Industry Expert Group. DG Information Society and Media – Directorate for Converged Networks and Service Manifesto (2009)
38. Nosratinia, A., Hunter, T., Hedayat, A.: Cooperative Communication in Wireless Networks. IEEE Comm. Mag. 42(10), 74–80 (2004)
39. Mitola III, J., Maguire Jr., G.: Cognitive Radio: Making Software Radios More Personal. IEEE Personal Comm. 6(4), 13–18 (1999)
40. Haykin, S.: Cognitive Radio: Brain-Empowered Wireless Communications. IEEE Journal of Selected Areas in Comm. 23(2), 201–220 (2005)
41. Haykin, S.: Cognitive Radio: Research Challenges. In: Vehicular Technology Conference Tutorial (2008)
42. Mitola III, J.: The Software Radio Architecture. IEEE Comm. Mag. 33(5), 26–38 (1995)
43. Hamdi, K., Letaief, K.: Cooperative Communications for Cognitive Radio Networks. Proceedings of the IEEE 97(5), 878–893 (2009)
44. Akyildiz, I., Lee, W., Vuran, M.: Next Generation/Dynamic Spectrum Access/Cognitive Radio Wireless Networks: A Survey. Computer Networks 50, 2127–2159 (2006)
45. Akyildiz, I., Lee, W., Vuran, M., Mohanty, S.: A Survey on Spectrum Management in Cognitive Radio Networks. IEEE Comm. Mag. 46(4), 40–48 (2008)
46. Lassila, P., Penttinen, A.: Survey on Performance Analysis of Cognitive Radio Networks. Helsinki University of Technology Project Report (2008)
47. Gershenfeld, N.R., Krikorian, R.D., Cohen, D.: The Internet of Things. Sci. Am. 291(4), 76–81 (2004)
48. Belpaire, A.: Internet of Things: Already a Reality Today. Eurescom Mess@ge Magazine 2, 10–11 (2009)
49. Herzog, U.: Wireless Sensor Networks: Perceive the World Like Never Before. Eurescom Mess@ge Magazine 3, 7 (2006)

50. Akyildiz, I.: A Survey in Sensor Networks. IEEE Comm. Mag. 40(8), 102–114 (2002)
51. AWISSENET, Analysis of AWSN Nodes/Platforms Security Holes. Ad-hoc Personal Area Network & WIreless Sensor SEcure NETwork (AWISSENET) Project Deliverable D2.2 (2008)
52. Zahariadis, T.: Trust Models for Sensor Networks. In: ELMAR, 50th International Symposium (2008)
53. AWISSENET, Trusted Path Discovery. AWISSENET Project Deliverable D3.2 (2008)
54. Hauswirth, M.: Global Sensor Networks: Enabling Networked Knowledge. Future Internet Assembly (FIA) Madrid, Spain (2008)
55. Bauer, M.: Bringing Context Information to the Real-World Internet. In: SENSEI uSWN Workshop presentation at Madrid, Spain (2008)
56. Egan, R.: Sensors & Ubiquitous Connectivity: Impact on Internet Network Architecture. In: First Japan-EU Symposium on NGN & FI (2008)
57. Rothenberg, C.E., Verdi, F.L., Magalhaes, M.: Towards a New Generation of Information-Oriented Internetworking Architectures. Re-Architecting the Internet, Madrid, Spain (2008)
58. PSIRP, Progress Report and Evaluation of Implemented Upper and Lower Layer. Publish-Subscribe Internet Routing Paradigm (PSIRP) Deliverable D2.3 (2009)
59. European Commission, Future Internet Assembly 2009, Conference Report (2009)
60. Marinescu, D., Kröger, R.: State of the Art in Autonomic Computing and Virtualization. Wiesbaden University of Applied Sciences Technical Report (2007)
61. Chowdhury, N., Boutaba, R.: A Survey of Network Virtualization. David R. Cheriton School of Comp. Science, University of Waterloo, Canada. Technical Report CS-2008-25 (2008)
62. Peterson, L., Shenker, S., Turner, J.: Overcoming the Internet Impasse through Virtualization. IEEE Computer 38(4), 34–41 (2005)
63. Carapinha, J., Jiménez, J.: Network Virtualization – a View from the Bottom. In: Proceedings of the 1st ACM workshop on Virtualized Infrastructure Systems and Architectures, pp. 73–80 (2009)
64. Touch, J., Hotz, S.: The X-Bone. In: 3rd Global Internet Mini-Conf. at IEEE Globecom, pp. 59–68 (1998)
65. Touch, J., Finn, G., Eggert, L., Hughes, A., Wang, Y.: The X-Bone & its Virtual Internet Architecture 10 Years Later. In: Workshop on Overlay and Net. Virtualization, 16th GI/ITG Conference on Kommunikation in Verteilten Systemen, Kassel, Germany (2009)
66. Peterson, L., Anderson, T., Culler, D., Roscoe, T.: A Blueprint for Introducing Disruptive Technology into the Internet. SIGCOMM Computer Comm. Review 33(1), 59–64 (2003)
67. White, B., Lepreau, J., Stoller, L., Ricci, R., Guruprasad, S., Newbold, M., Hibler, M., Barb, C., Joglekar, A.: An Integrated Experimental Environment for Distributed Systems and Networks. In: Proc. of the 5th Symposium on Operating System Design and Implementation, Boston, USA (2002)
68. GENI, GENI Design Principles. Computer 39(9), 102–105
69. Bavier, A., Feamster, N., Huang, M., Peterson, L., Rexford, J.: In VINI Veritas: Realistic and Controlled Network Experimentation. In: Proc. of SIGCOMM 2006, New York, USA (2006)

70. Raychaudhuri, D., Seskar, I., Ott, M., Ganu, S., Ramachandran, K., Kremo, H., Siracusa, R., Liu, H., Singh, M.: Overview of the ORBIT Radio Grid Testbed for Evaluation of Next-Generation Wireless Network Protocols. In: Proc. of IEEE Wireless Comm. and Net. Conference (2005)
71. Mahindra, R., Bhanage, G., Hadjichristofi, G., Seskar, I., Raychaudhuri, D.: Space Versus Time Separation For Wireless Virtualization on An Indoor Grid. In: Proc. of IEEE NGI (2008)
72. Peterson, L., Anderson, T., Culler, D., Roscoe, T.: A Blueprint for Introducing Disruptive Technology into the Internet. In: Proceedings of the 1st ACM Workshop on Hot Topics in Networks, Princeton, New Jersey, USA (2002)
73. Dempsey, H.: GENI: Global Environment for Network Innovations. In: Future of the Internet Conference, Bled, Slovenia (2008)
74. Völker, L., Martin, D., Khayat, I., Werle, C., Zitterbart, M.: A Node Architecture for 1000 Future Networks. In: IEEE Future-Net 2009 (2009)
75. Haider, A., Potter, R., Nakao, A.: Challenges in Resource Allocation in Network Virtualization. In: 20th ITC Specialist Seminar, Hoi An, Vietnam (2009)
76. Ramaswami, R., Sivarajan, K.: Optical Networks: A Practical Perspective. Morgan Kaufmann, San Francisco (1998)
77. Poikselka, M., Mayer, G.: The IMS: IP Multimedia Concepts and Services. Wiley, Chichester (2009)
78. Morita, L.: NCSL, Next Generation Network Standards in ITU-T. In: Broadband Convergence Networks, Vancouver, Canada (2006)
79. Shrestha, S.L., Lee, J., Chong, S.: Virtualization and Slicing of Wireless Mesh Network. In: 3rd Int. Conf. on Future Internet Technologies, Seoul, Korea (2008)
80. Niebert, N., El Khayat, I., Baucke, S., Keller, R., Rembarz, R., Sachs, J.: Network Virtualization: A Viable Path Towards the Future Internet. Wireless Personal Comm 45, 511–520 (2008)
81. Söllner, M., Görg, C., Pentikousis, K., Lopez, J.C., Leon, M.P., Bertin, P.: Mobility Scenarios for Future Internet: The 4WARD Approach. In: The 11th International Symp. Wireless Personal Multimedia Comm. (2008)
82. Smith, G., Chaturvedi, A., Mishra, A., Banerjee, S.: Wireless Virtualization on Commodity 802.11 Hardware. In: International Conf. on Mobile Comp. and Net., pp. 75–82 (2007)
83. GENI, Technical Document on Wireless Virtualization. Global Environment for Network Innovations (GENI) Technical Report GDD-06-17 (2006)
84. Yap, K., Kobayashi, M., Underhill, D., Seetharaman, S., Kazemian, P., McKeown, N.: The Stanford OpenRoads Deployment. In: WiNTECH Mobicom, Beijing, China (2009)
85. Nissenbaum, H.: The Meaning of Anonymity in an Information Age. The Information Society 15, 141–144 (1999)
86. Sullivan, K.: On the Anonymity "versus" Accountability Debate. Risk Manager magazine, Ireland (2009)
87. Hovav, A., Berger, R.: Tutorial: Identity Management Systems and Secured Access Control. Communications of the Association for Information Systems 25: Article 42 (2009)
88. Kephart, J.O., Chess, D.M.: The Vision of Autonomic Computing. IEEE Computer Magazine 36(1), 41–50 (2003)
89. McCann, J., Huebscher, M.: A Survey of Autonomic Computing - Degrees, Models and Applications. ACM Computing Surveys 40(3) (2008)

90. Dobson, S., Sterritt, R., Nixon, P., Hinchey, M.: Fulfilling the Vision of Autonomic Computing. Computer Magazine 43(1), 35–41 (2010)
91. Clark, D., Partridge, C., Ramming, J., Wroclawski, J.: A Knowledge Plane for the Internet. In: Proc. of the Conference on Applications, Technologies, Architectures, and Protocols for Computer Comm., Karlsruhe, Germany (2003)
92. Smirnov, M.: Autonomic Communication: Research Agenda for a New Communications Paradigm. Fraunhofer FOKUS technical Report (2004)
93. Zseby, T., Hirsch, T., Kleis, M., Popescu-Zeletin, R.: Towards a Future Internet: Node Collaboration for Autonomic Communication. Towards the Future Internet. IOS Press, Amsterdam (2009)
94. Dobson, S., Denazis, S., Fernández, A., Gaïti, D., Gelenbe, E., Massacci, F., Nixon, P., Saffre, F., Schmidt, N., Zambonelli, F.: A Survey of Autonomic Communications. ACM Transactions on Autonomous and Adaptive Systems 1(2), 223–259 (2006)
95. Galis, A., Denazis, S., Bassi, A., et al.: Management Architecture and Systems for Future Internet Networks. Towards the Future Internet. IOS Press, Amsterdam (2009)
96. Chaparadza, R., Papavassiliou, S., Kastrinogiannis, T., Vigoureux, M., Dotaro, E., Davy, A., Quinn, K., Wodczak, M., Toth, A.: Creating a Viable Evolution Path Towards Self-Managing Future Internet via a Standardizable Reference Model for Autonomic Network Engineering. In: Future Internet Assembly (FIA) Prague, Czech Republic (2009)
97. Pentikousis, K., Meirosu, C., Miron, A., Brunner, M.: Self-management for a Network of Information. In: IEEE International Conf. on Comm., Dresden, Germany (2009)
98. Kaloxylos, A., Rosowski, T., Tsagkaris, K., Gebert, J., Bogenfeld, E., Magdalinos, P., Galani, A., Nolte, K.: The E3 Architecture for Future Cognitive Mobile Networks. In: IEEE 20th International Symposium on Personal, Indoor and Mobile Radio Communications, Tokyo, Japan (2009)
99. Smirnov, M., Tiemann, J., Chaparadza, R., Rebahi, Y., et al.: Demystifying Self-awareness of Autonomic Systems. In: ICT-MobileSummit, Santander, Spain (2009)
100. FET, Shaping the Future: Report on FET Consultations 2007-2008. Future and Emerging Technologies (FET) Proactive, European Commission Report (2008)
101. Altman, E., Dini, P., Miorandi, D., Schreckling, D., et al.: Paradigms for Biologically-Inspired Autonomic Networks and Services. The BIONETS Project eBook (2010)
102. Schmidhuber, H.: Ultimate Cognition à la Gödel. Cognitive Computation 1(2), 177–193 (2009)
103. Legg, S.: Machine Super Intelligence. Faculty of Informatics of the University of Lugano Doctoral Dissertation (2008)
104. Wang, Y.: On Cognitive Informatics. In: Proc. 1st IEEE International Conference on Cognitive Informatics, Calgary, Canada (2002)
105. Wang, Y.: On Cognitive Computing. Int. J. of Software Science and Computational Intelligence 1(3), 1–15 (2009)
106. Jacobson, V., Smetters, D., Thornton, J., Plass, M., Briggs, N., Braynard, R.: Networking Named Content. In: CoNEXT 2009, Rome, Italy (2009)
107. Ahlgren, B., D'Ambrosio, M., Dannewitz, C., Marchisio, M., Marsh, I., Ohlman, B., Pentikousis, K., Rembarz, R., Strandberg, O., Vercellone, V.: Design Considerations for a Network of Information. In: Re-Architecting the Internet, Madrid, Spain (2008)
108. Tarkoma, S., Ain, M., Visala, K.: The Publish/Subscribe Internet Routing Paradigm (PSIRP): Designing the Future Internet Architecture. Towards the Future Internet. IOS Press, Amsterdam (2009)

109. Ohlman, B., Ahlgren, B., et al.: Networking of Information: An Information-centric Approach to the Network of Future. In: ETSI Future Network Technologies Workshop (2010)
110. Niebert, N.: Vision on Future Content Networks: A Networks and Media Joint Venture. In: Future Internet Assembly (FIA), Madrid, Spain (2008)
111. Kopecký, J., Domingue, J., Fensel, D., González-Cabero, R.: SOA4All, Enabling the SOA Revolution on a World Wide Scale. In: 2nd IEEE International Conf. on Semantic Computing Santa Clara, USA (2008)
112. Berners-Lee, T., Hendler, J., Lassila, O.: The Semantic Web. Scientific American Magazine 23(1) (1999)
113. Cross-ETP, Future Internet Strategic Research Agenda. Future Internet X-ETP Group Technical Report v1.1 (2010)
114. Paulson, L.D.: News Briefs - W3C Works on Semantic Web Proposal. Computer Magazine 36(11), 20 (2003)
115. Fensel, D.: ServiceWeb 3.0. In: IEEE/WIC/ACM International Conf. on Intelligent Agent Technology, Fremont, USA (2007)
116. Krummenacher, R., Norton, B., Simperl, E., Pedrinaci, C.: SOA4All: Enabling Webscale Service Economies. In: 3rd IEEE International Conf. on Semantic Computing, Berkeley, USA (2009)
117. 4WARD, Architecture and Design for the Future Internet: Second NetInf Architecture Description. Deliverable D6.2 (2010)
118. Eugster, G., Felber, P., Guerraoui, R., Kermarrec, A.M.: The Many Faces of Publish/Subscribe. ACM Computing Surveys 35(2), 114–131 (2003)
119. Smetters, D.K., Jacobson, V.: Securing Network Content. PARC Technical Report (2009)
120. MIT, Identity in an Information-Centric Internet. MIT Privacy and Security Working Group Technical Report (2008)
121. Gruber, T.: A Translation Approach to Portable Ontology Specifications. Knowl. Acquis. 5(2), 199–220 (1993)
122. Schmidt, A., Pasic, A., Martinelli, F., Le Métayer, D., Waller, A., Ivanov, I., Dooly, Z.: Security Challenges in Future Internet. Networked European Software and Services Initiative Position Paper (2008)
123. Clark, D.D., Sollins, K.R., Wroclawski, J., Braden, R.: Tussle in Cyberspace: Defining Tomorrow's Internet. IEEE/ACM Trans. on Networking 13, 462–475 (2005)
124. Bush, R., Meyer, D.: Some Internet Architectural Guidelines and Philosophy. IETF RFC 3439 (2002)
125. Papazoglou, M., Traverso, P., Dusdar, S., Leymann, F.: Service-Oriented Computing Research Roadmap. Technical Report/Vision (2006)
126. Nitto, E., Karastoyanova, D., Metzger, A., Parkin, M., Pistore, M., Klaus, P., Silvestri, F., Van den Heuvel, W.: S-Cube: Addressing Multidisciplinary Research Challenges for the Internet of Services. Towards the Future Internet. IOS Press, Amsterdam (2009)
127. Pistore, M., Traverso, P., Paolucci, M., Wagner, M.: From Service Services to a Future Internet of Services. Towards the Future Internet. IOS Press, Amsterdam (2009)
128. Erl, T.: Service-Oriented Architecture. Prentice-Hall, Englewood Cliffs (2004)
129. Tselentis, G., Domingue, T., Galis, A., Gavras, A., Hausheer, D., Krco, S., Lotz, V., Zahariadis, T., et al.: Towards the Future Internet: A European Research Perspective. IOS Press, Amsterdam (2009)

130. S-Cube Consortium, Comprehensive Overview of the State of Art on Service-Based Systems. Deliverable # CD-IA-1.1.1 S-Cube Project (2008)
131. EIFFEL Think-Tank Consortium, Starting the Discussion. EIFFEL Report (2009)
132. Trossen, D.: Invigorating the Future Internet Debate. ACM SIGCOMM Computer Comm. Review 39(5), 44–51 (2009)

New Generation Internet Architectures: Recent and Ongoing Projects

Tania Regina Tronco[1], Takashi Tome[1], Christian E. Rothenberg[2], and Antonio Marcos Alberti[3]

[1] CPqD Foundation, Rodovia Campinas Mogi-Mirim, km 118,5,
Campinas – São Paulo, CEP 13096-902, Brazil
{tania,takashi}@cpqd.com.br
[2] University of Campinas (UNICAMP), Cidade Universitária "Zeferino Vaz"
Distrito de Barão Geraldo - Campinas - São Paulo, CEP 13083-852, Brazil
chesteve@dca.fee.unicamp.br
[3] Instituto Nacional de Telecomunicações (INATEL), Av. João de Camargo 510, Santa Rita do Sapucaí, Minas Gerais, Brazil, CEP 37540-000,
alberti@inatel.br

Abstract. New generation Internet architectures projects are popping up everywhere with new designs proposals and protocols. It is time to rethink the Internet architecture and reengineering it to address the current and future requirements. This text survey recent and ongoing projects focusing on three driving scenarios for the future Internet: object-centric, content-centric and user-centric. An overview about the future Internet research activities in U.S., Europe, Japan and Brazil is also presented.

1 Introduction

The Internet has invaded most aspects of life and society, changing our lifestyle, work, communication and social interactions and giving us unprecedented expectations about new forms of interactivity with the surroundings, access to global knowledge, and decrease of the digital divide. Nevertheless, the current Internet suffers with lack of mobility, loss of transparency, scalability problems, incompatibility issues, security vulnerability and attacks, mainly due to protocols taking roles for which they were not originally designed security vulnerability and attacks. As a consequence, a big momentum on Future Internet (FI) research has emerged; it is time to rethink the Internet architecture and reengineering it to address the current and future requirements. There is a common consensus that the Internet needs improvement. Nevertheless, there is not yet a shared vision on how this may happen. There is not a complete network science to accurately predict and control network behaviors with global interactions. New theories and methodologies are being developed to help understanding this planet-scale complex system.

The Internet architecture reengineering research includes: (i) rethinking its fundamentals principles and give them coherence in accordance with new requirements towards a global information infrastructure, (ii) experimentally-driven research for validation of the new proposals at scale and under realistic scenarios and (iii) business and social incentives for adoption.

Diverse approaches and visionary ideas have emerged and there are two main approaches towards changing Internet architecture:

- **Incremental approach:** Internet architectural changes are effected by adding new functionalities and protocols to the current architecture;
- **Clean-Slate approach:** redesign the Internet architecture from the scratch with the current network knowledge and aim at the development of more intelligent and adaptive solutions to achieve better resource utilization, power saving without the limitations of the current architectural design. Clean slate design is a more free thinking, not presuming a clean slate deployment.

Future Internet research projects are popping up everywhere with new architecture designs and protocols. In general, the current ongoing Future Internet projects can be aggregated in two different groups: (i) exploratory research of new reference architecture and (ii) experimental facilities.

The first group includes new architectures designs, such as:

- Overlay networks
- New control and management architectures
- Network virtualization
- Locator-identifier split
- Information-oriented networks
- User-centric networks
- Internet of things
- Security, privacy and trust
- Internet of services
- Substrate networking
- Revisiting networking fundamentals

The experimental facilities, the second group, provide experimentation services to the future Internet research community at scale.

In the U.S., future Internet research activities started at the end of 2005 when National Science Foundation (NSF) launched the Future InterNet Design (FIND) research program. FIND is within NSF NeTS program and search the requirements for a global network of 15 years from now [1]. In 2006, 26 research projects received resources via NetSE. In 2007, this number was increased to 54 projects. The main new requirements identified for the Internet architecture were:

- Security;
- Reliability and availability;

- Better management;
- Support to future applications;
- Use future network technologies;
- Achieve social needs;
- Long life.

This program motivated the development of a Global Environment for Network Innovations (GENI) Project to solve the difficulties in doing research with a network in operation. The established network infrastructure does not allow the research community to create new (radical) solutions to solve existing problems, neither stop the operational infrastructure for testing new protocols proposals. Diverse workshops took places with discussions about how to build a new infrastructure capable of supporting the development of new network solutions. Finally, NSF concluded that was important and necessary to finance the building of a new infrastructure named GENI to support experimentation at scale and in real time, enabling a large number of experiments in parallel through virtualization process [2].

The GENI infrastructure is composed by a high-capacity optical network, a programmable and federated core, large clusters of CPUs and disks, diverse types of wireless access technologies and sensor networks. The experiments are scheduled and run independently of each other using programmable components via an end-to-end virtualized slice [3]. The virtualization process is implemented by software-defined networking technology, e.g. OpenFlow, which is being developed by a research group at Stanford University [4]. Design, prototyping and construction of GENI are performed by the research community with a special is emphasis on openness using virtualization.

In Europe, research activities are mainly under multi-year continent-wide Framework Programme (FP), which cover a wide range of subjects, from ICT to energy, nanotechnology, health, etc. Current programme is the seventh (FP7), started in January 2007 and will expire in 2013, embodies the following research clusters:

- FCN (Future Content Networks)
- FISO (Future Internet Service Offer)
- MANA (Management and Service-aware Networking Architectures)
- FIRE (Future Internet Research & Experimentation)
- FISE (Future Internet Socio-Economics)
- RWI (Real World Internet)
- TI (Trust and Identity)

FCN claims that the Future Internet will be centered in content and its treatment. It proposes two content-centric architectures for FI: (i) an evolutionary architecture, where virtual nodes are hierarchically organized over a substrate infra-structure to establish content/service aware virtual clouds and overlays; and other (ii) clean–slate designed, where autonomous content objects are hierarchically organized, divided, combined and transported over the network. The idea is to create content experiences for users using autonomic service/application and content objects combination. While the first architecture relies on virtualization, the second one relies on autonomic and content-centric approaches.

FISO aims to promote service-based software interfaces to integrate and interwork FI overlay software components. The project cluster investigated different scenarios to determine service role on FI architecture. FISO relays on service oriented design paradigm to address service life cycling, service level agreement, service contextualization, service reference models and architecture.

MANA focus on architectures to create manageable service-aware networks and network-aware service platforms for FI. MANA provides a research orientation that covers the following capabilities: infrastructure, control, elasticity, accountability, virtualization, self-management, service enablement and orchestration. MANA defines an architectural model with four types of interfaces [5]: (i) α-interfaces to enable service and applications development; (ii) β-interfaces to provide service-aware orchestration of virtual resources; (ii) γ-interfaces to set up virtualization systems through network programmability and self-management; and (iv) δ-interfaces to provide access to substrate resources. The proposed model relays on programmability, virtualization, autonomic ICT and many *-centric approaches. It proposes virtualization not only of networks and nodes, but also of data and service centers. Service availability, ubiquitous connectivity and mobility, anywhere, anytime, are also concerns. Network elements must implement autonomous control loops to provide a vast list of self-functionality, such as self-stability, self-configuration, self-optimization, self-healing, among others. Other concerns are service description, discovery, negotiation, management and service life cycle, which takes different approaches depending on related interface (α,β,γ or δ). Self-* properties are envisioned even for services. MANA also proposes the orchestration of software systems to improve dynamically architecture behavior accordingly to expected goals, policies and business processes. Finally, MANA concerns to unite many *-centric view points, such as management-centric, information-centric, context-centric, content-centric, object-centric, etc.

FIRE promotes the integration of the FP7 projects in an environment for investigation and experimentation of new (evolutionary and clean-slate) paradigms. It includes experimental facilities such as: OneLab, Panlab, Federica, WISEBED and VITAL++ and experimentally-driven, multi-disciplinary research such as:

- SelfNet (Self-Management of Cognitive Future InterNET Elements);
- SmartNet (SMART-antenna multimode wireless mesh Network);
- ECODE (Experimental COgnitive Distributed Engine);
- OPNEX (Optimization driven Multi-Hop Network Design and Experimentation);
- Nanodatacenters;
- ResumeNet (Resilience and Survivability for Future Networking: Framework, Mechanisms, and Experimental Evaluation) and
- N4C (Networking for communications challenged communities: architecture, test beds and innovative alliances).

RWI defends the notion that real world devices could provide useful information for the Internet, enriching software services and applications with contextualized information. Ubiquitous devices with Internet access allow creating the so called ubiquitous connectivity, i.e. connectivity anywhere, anytime, for anything and everyone. RWI believes that in the near future trillions of devices will acquire real world information to provide real time connection between reality and virtual worlds. The role of network embedded devices goes far beyond connectivity, advancing to the collaboration to achieve common goals, to information integration and contextualization, to content generation, and even to create what RWI defines as "social devices", which are able of interact each other to exchange knowledgement about situations and potential problems. RWI defends that devices are uniquely identified. The FIA working group suggests some guidelines for research in the area, including management, scalability, heterogeneity, contextualization, knowledge exchanging, privacy, security and trust.

TI working group focuses on trust and identity challenges in FI taking a cross-domain approach and the main concerns are: (i) provisioning of electronic Identity (eID) for humans and other entities; (ii) eIDs management and governance; (iii) mechanisms to increase trustworthiness and measure trust and security; (iv) scalability and scope of IDs, trust and privacy mechanisms; (v) the accountability and privacy debate. TI is working to establish an identity management framework as well as to design architectural aspects for trust and identity.

The recently approved PPP (Public-Private Partnerships) are meant to complements the FP7 activities bridging the gap between technologies and key applications sectors e.g. telecommunication, energy, health and transport. ICT research is also promoted by European Commission program on Future and Emerging Technologies (FET).

FP7 also promotes the FIA (Future Internet Assembly), where FP7 participants meet twice a year e.g. Stockholm in November 2009 and Valencia in April 2010.

In Japan, the AKARI project, sponsored by the National Institute of Information and Communications Technology (NICT) has a working group on development of a new network architecture following the clean-slate approach towards a NeW Generation Network (NWGN) by 2015 [6]. The NWGN idea started in Japan at end 2007 within the NWGN Forum.

NICT's Vision for NWGN it to maximize the potential to innovation, cultural diversity, knowledge society, productivity, quality of life, human wisdom and minimize the negative points such as energy issues, inequality, medical issues, food issues and aging society with few children.

The network architecture proposal is based on five network targets [7] as follows:

- Value Creation Network
    o Service creation network
    o Media creation network
- Trustable Network
    o Social infrastructure for trustable network
    o Trustable networks for human and society

- Ambient/Ubiquitous Network
    - Global-scale sensor cloud
    - Surrounding network
- Self-Management Network
    - Network for diversity
    - Network unification
- Sustainable Network
    - Green Network
    - Dynamic spectral resource management

NWGN design principles for creating new generation network architecture are KISS (Keep It Simple, Stupid), Sustainable and Evolutionary and Reality Connection [6]. The idea is to determine promising technologies to design architecture, integrate and simplificate them. The architecture must be evolvable and sustainable, to allow evolution without replacement in succeeding decades. The design considers self-organization, self-emergent behavior, and other self-* properties. Distributed control is the key to deal with network scalability. The design must be robust to deal with large scale simultaneous failures and attacks. Mobility is also an issue as well as topology fluctuations. The network must provide real-time measurements to achieve better controllability. Reality connection means to relate NWGN functioning with real-world society in order to achieve better security, accountability, privacy, etc.

The main components of the new generation network architecture are optical packet switching and optical paths, optical and wireless access, a transport layer control, identifier/locator split principle and network virtualization. Proposed architecture has three layers: underlay, common and overlay. The underlay layer provides high capacity, secure, ubiquitous, scale free and stable connectivity. It aims to achieve global self-emergent behavior by means of autonomic elements. The common layer will replace IP and provides flexibility, quick control and cross layer mechanisms to the other two layers. It is a mediation layer. The overlay layer provides adaptability by means of evolutionary and customizable overlay networks. It relays on autonomic operation of virtualized resources. Experimentation efforts are taking place at Japan Gigabit Network Plus (JGN2plus) and Network Virtualization and Overlay Network Research Laboratory (NVLab).

In Brazil, FUNTTEL (Brazilian Funding for Technological Development of the Telecommunications) of Ministry of Communications supports exploratory and experimental projects in R&D institutions, industry and telecom operators to catalyze discoveries and innovation in the Future Internet such as:

- ARCMIP (ARchitectures for Mobile IP Project) project [8], started in July, 2008, aims at exploring new network architecture designs and to identify research challenges presenting ambitious to:
    - Development of telecommunications products for national industry;
    - Applications on public services and
    - Maintain Brazil aligned with international research efforts shaping a long term research (joint) agenda.

New Generation Internet Architectures: Recent and Ongoing Projects

- GIGA Project [9], started in 2003, is a high speed IP/WDM network. It was the first large-scale experimental network in South America and connects diverse universities, research centers and telecom operators in Brazil.
- Horizon Project [10], started in 2009, is a bi-national research project selected by the Brazilian Communication Ministry and the French National Research Agency (ANR).

Also in Brazil, the Kyatera network [11] was created to gather together competences and laboratory resources to develop science, technologies, and applications of the future Internet.

RNP is Brazil´s NREN (National Research and Education Network) [12], fully supported by the federal government to provide advanced network services to the higher education and research community.

In the following sections, we survey recent and ongoing projects focusing on three driving scenarios for the Internet: object-centric, content-centric and user-centric scenarios, as identified by the ARCMIP project. The objective here is not to present an exhaustive list of these projects, but those with relevant features to the formulation of new architectural proposals according to ARCMIP's point of view.

2 Recent and Ongoing Projects on User-Centric Scenario

The user-centric scenario is related to provide a ubiquitous and comfortable services portfolio for the people and by people with the rationale of changing the focus of the Internet to prioritize users' needs.

This new approach has resulted in a series of project proposals to provide properties of "consciousness" for the network achieves users' requirements such as personal preferences, location, context, self-servicing, usability, quality of experience, etc. Some research topics directly related to the user-centric scenario includes:

- Network awareness regarding user environment;
- Location-based services - Internet of services;
- Service oriented design, compose-ability and orchestration;
- Inclusion of self-* and *-aware properties, to create autonomic networks with self-management functionalities (self-configuration, self-healing, self-protection and self-optimization) and increasing self-awareness and situation-awareness;
- QoS-aware networks: routing protocols with restrictions, differentiation of traffic flows, traffic engineering, fairness in resource usage, etc;
- Virtualization: allowing the creation of various types of overlay customized virtual networks over substrate networks to fit the variety of services and user requirements;
- Security, privacy and trust for users, applications and services;
- New communication models, such as consented, disruptive, etc.

Some recent and ongoing projects include:

- Your Way
- Daidalos
- C-Cast
- CHIANTI
- 4WARD
- ANA
- AutoI
- SOA4All

2.1 YourWay

YourWay [13] is a bilateral three-year project between the Service-Oriented Research Unit of FBK-Irst, Trento, Italy and DoCoMo Euro-Labs, Munich, Germany started in 2007. The project aims the service composition as a new paradigm for user-centric service provisioning in mobile environments. As a practical result, a prototype implementation of a platform for user-centric composition of mobile services was implemented.

2.2 Daidalos

Daidalos (Designing Advanced network Interfaces for the Delivery and Administration of Location independent, Optimized personal Services) [14] is a project supported by FP6 in the period from 2008 to 2009. It uses a user-centric approach guided by concepts such as mobility management, AAA, resource management, virtual identity, ubiquitous and seamless pervasiveness, integrates broadcast technologies and federation among different players (dynamic business environment).

2.3 C-Cast

C-Cast (Context-Casting) [15] is a project supported by FP7 in the period from 2008 to 2010 that aims to research, design and develop context and group management service enablers to support context casting applications and services of mobile multicast context aware services.

2.4 CHIANTI

CHIANTI [16] is a project supported by FP7 in the period from 2008 to 2009 that follows a user-driven approach improving disconnection and disruption tolerance for mobile user communications by deploying a new service-support infrastructure operated by a third party as an overlay network.

2.5 4WARD

4WARD [16] is a project supported by FP7 in the period from 2008 to 2009 that aims to improve the quality of life for European citizens by creating a family of dependable and interoperable networks providing direct and ubiquitous access to information. It explores a new approach to allow for a plurality and multitude of network architectures via network virtualization. It also uses in-network management functionally that is a new paradigm for network management, where management functions come as embedded capabilities of the devices.

2.6 ANA

ANA (Autonomic Network Architecture) [16] is a European project started in January, 2006 and finished in December, 2009 within IST – FET (Future Emerging Technologies) program. It aims to identify fundamental autonomic network principles and design and develop a novel autonomic network architecture that enables flexible, dynamic, and fully autonomous formation of network nodes as well as whole networks.

2.7 AutoI

AutoI (Autonomic Internet) [16] is a FP7 project started in January, 2008 and finished in December, 2009. It aims to conceive new in-network management architecture for Future Internet (FI). The idea is to self-manage virtualized resources in order to achieve mobility, security, quality and reliability. AutoI is composed by five distributed systems (or planes), namely: Orchestration, Service Enablers, Knowledge, Management and Virtualization (OSKMV). These systems collect operational information and based on rules generate *-aware actions.

2.8 SOA4All

SOA4All [16] is a project supported by FP7 in the period from 2008 to 2011 that aims to provide a framework and infrastructure to integrate semantic Web and context management into SOA (Service Oriented Architecture).

3 Ongoing Projects - Object-Centric Scenario

The object-centric scenario opens the Internet scale connectivity to any imaginable real word object, expanding host and device endpoints spaces to sensors and things. There are several recent and ongoing projects worldwide on the Internet of the things itself, and especially related to technologies that can be incorporated on it such as:

- IrisNet
- Hourglass
- e-SENSE
- Ubiquitous Sensor Networks
- SENSEI
- SENDORA
- AWISSENET

3.1 IrisNet

IrisNet (Internet-scale Resource-Intensive Sensor Network Service) [17] is a joint project between Intel, Carnegie Mellon University (CMU) and University of Berkeley, having its activities peak between 2002 and 2005.

The project goal was to develop a structured and simple model to treat a sensor network as a large database based on XML. Although a generic model, it was designed to use cameras (generating flows of images) as sensors.

3.2 Hourglass

Hourglass [18] was a project of Harvard University to create a universal infrastructure for sensor networks, with activities peak in the period 2004-2005.

One of the results was a novel mechanism to deal with the occasional disconnections that occur in sensor networks. This mechanism monitors the links sending/receiving "heart beat" messages or using data traffic analysis. When a disconnection is detected, the data are stored in buffers.

One of its shortcomings was scalability to keep explicit information about the connection status of each node [19].

3.3 e-SENSE

e-SENSE was a project supported by FP6 in the period from January 2006 to 31st of December 2007. The project goal was the capture of ambient intelligence through Wireless Sensor Networks and to integrate it in the IMS (IP Multimedia Subsystem) architecture. e-SENSE allows IMS applications to collect information about the environment using diverse types of WSAN networks.

The e-SENSE architecture consists of two main components. The first is the service enabler, which provides information about the context from information collected by sensors. This information is available through standard services of IMS. With that, the context service enabler can be used as a building block to create various context based services in an IMS environment. The second component is a gateway located between the IMS networks and the WSANs. The communication between the WSANs and the gateway is based on publish/subscribe paradigm [19].

3.4 Ubiquitous Sensor Networks

The Ubiquitous Sensor Networks project, led by Telefonica (Spain), has the same goal of e-SENSE, i.e., the integration of sensor networks to IMS. The internal mechanisms are also similar, as well the communication with the gateway based publish/subscribe.

3.5 SENSEI

SENSEI project is part of the FP7, with 19 organizations participants, from which 11 European countries and a budget of 23 million for 3 years (2008-2010). Conceptually, the project is derived from e-Sense [20].

The project goal is to create SENSEI architecture (framework) and a pluggable and global wireless network of sensors and actuators that meet certain requirements for scalability and reliability. Additionally, this architecture will enable the integration of heterogeneous WSANs islands that are currently disconnected. The project's vision an integrated environment, so that any sensor, actuator or service is accessible through a universal interface. Additionally, in the same network will coexist different types of sensors with diverse traffic patterns. The universal interface should enable the creation of new services and applications to run on the existing network. Therefore, the main idea is to translate the benefits of a universal Internet and the programmability of computers for an Internet of the objects.

SENSEI [19] examines various proposals for sensor networks and concludes that its main weaknesses are:

- Lack of semantics and ontology mechanisms for a more complex processing of information and to support interactive applications with a high degree of abstraction;
- Limited support for mediation, requiring, at least in some cases, a detailed knowledge of the sensor networks or specific services specification;
- Lack of scalability in case of using centralized control architectures;
- The complexity of the mechanisms for connection management, inadequate to treat short-time connections and event-based interactions;
- Lack of support for composition of distributed services;
- Lack of support for accounting and auditing (accountability);
- Lack of mechanisms to access control among interactions between services and WSANs as well to address reliability (trust), privacy and information provided by WSANs;
- Does not address the issues of mobility and the sudden unavailability of service in case of long time communications;
- Does not provide support to ensure the QoI (Quality of Information) and QoA (Quality of Actuation) a service;
- Has no support for arbitrary allocation of network resources (necessary for actuators sensors);

- Inability to dynamically adapt to context changes;
- Lack of support for more complex services creation.

Nowadays, this is an intense research area at the international level, needing a continuous monitoring to observe the development of new proposals.

3.6 SENDORA

The FP7 Sensor Network for Dynamic and Cognitive Radio Access [16] project started in January 2008 and will finish in December 2010. The project uses sensor nodes to monitor spectrum holes for a cognitive radio network. Detected opportunities provide real world based dynamic spectrum allocation. SENDORA also covers business cases for proposed solution as well as proposes a reconfigurable architecture for it. In such architecture, the cognitive radio network send frequency opportunity queries for the wireless sensor network, which returns reports on frequency holes availability.

3.7 AWISSENET

The Ad-hoc personal area network and WIreless Sensor SEcure NETwork [16] FP7 project focuses on securing such networks against threads ranging to physical jamming up to false information sinks. The project presents the state-of-art in Wireless Sensor Networks (WSNs) security, and analyzes how such networks constraints affect it. AWISSENET defends trust concept as a key approach to provide secure information routing in WSNs. It also covers secure service discovery and intrusion detection.

4 Ongoing Projects - Content-Centric Scenario

There are several projects worldwide to define new architectures for the Future Internet as an "information network". The base of this approach is that the Internet architecture would be radically different if it were designed with the current focus in mind on accessing and sharing identifiable pieces of content, and considering the current technology trends, where the costs of memory and processing are reducing faster than bandwidth costs. A series of projects funded by agencies of both sides of the Atlantic share the motivation to change the view and search for new network architectures for the Internet of the Future but the proposals differ, depending on the mechanisms and protocols developed. Next, diverse proposals based on decoupling (spatial and temporal) of endpoints of communication: sender and receiver, especially those who consider the content object as the first class object in the new Internet are described, such as:

- I3 (Internet Indirection Infrastructure)
- LNA (Layered Naming Architecture)

- TRIAD (Translating Relaying Internet Architecture)
- DONA (Data-Oriented Network Architecture)
- PSIRP (Publish / Subscribe Internet Routing Paradigm)
- CCN (Content-Centric Networking)
- Haggle
- Postcards from the Edge
- SCAFFOLD (Service-Centric Architecture For Flexible Object Localization and Distribution)

4.1 I3 - Internet Indirection Infrastructure

I3 [21] is project of University of California, Berkeley started in 2002 and finished in 2005 that aims at facilitating the development of multicast, anycast and mobility services. The proposal is based on a rendezvous mechanism that decouples the sending act from the receiving act. In I3, hosts are associated with identifiers, which are stored in the network as triggers. Triggers are composed by (*id*, *addr*) to indicate that all packets with an identifier *id* should be sent by the I3 network to the host located at *addr*.

Based on this model, the creation of a multicast group is equivalent to making all participants of a group register triggers with the same identifier.

In I3, mobile stations can maintain connectivity updating triggers when the IP address change. I3 consists of a set of servers to store triggers and forward packets (using IP) to other servers and I3 end systems. When a host wants to send a packet, it forwards it to one known server and if it does not contain the desired trigger, the packet is forwarded to other servers until the packet reaches it. I3 provides a best effort service as the current Internet, and does not implement reliability over the IP network.

4.2 Layered Naming Architecture (LNA)

LNA [22] work was done as part of the IRIS project, supported by the National Science Foundation. It proposes a semantic division between information identification (Service Identifier - SID) and its location (Endpoint Identifier - EID). In this architecture, the user only needs to know the SID of the service he wants to have access, independently of his current location (EID).

Such architecture requires the introduction of three additional levels to resolve identifiers: (i) first level to convert information from user level identifier to service identifier and create the SID, (ii) second level to find the endpoint identifier (EID) to that service identifier (SID) and (iii) third level to convert EID to IP address.

Some LNA advantages are:

- SIDs solves the problem of using URLs to name data and services and tie them to an endpoint;
- Using SIDs, the applications are named permanently, regardless of their location and treating services and data objects as the "first class" objects;

- Naming hosts using EIDs provides a natural solution for mobility and multi-homing: If a station identified by an EID changes the IP address, the EID resolution layer is in charge of updating the new IP address. This functionality enables automatic, continuous operation in the presence of mobility providing multi-homing in case of network path failures.
- Finally, it allows the natural addition of middleboxes such as NAT and Firewalls without violate end-to-end principle of the Internet architecture.

The challenges of implementing the LNA include:

- SIDs lack of a hierarchical domain-like structure; being a (randomly-looking) sequence of bits, or numbers that need DHT (Distributed Hash Table)-like mechanisms to enable an efficient resolution process;
- SIDs represents the address of a service or information; the user may wish to retain it, as it does currently with a URL. However, a SID gives no mnemonic association, which makes it unsuitable for users;
- LNA architecture adoption requires modifications in the transport and application layers to include the new layers for name and address resolution, while preserving the old architecture in operation. This implementation is not simple, but can occur incrementally and changes in applications and operating systems are needed to maintain compatibility with the legacy, allowing a smooth transition.

4.3 TRIAD

TRIAD [23] was a project of Stanford University, California, during the period from 1999 to 2004. It proposes a content routing primitive based on forwarding packets upon names, not in IP address. Users do not need to request for connectivity to a particular server or IP address, only request for a content items specified by their name (usually a URL).

In order to perform this type of routing, the Internet core should maintain and distribute information about the accessibility to content's items and routers must implement new methods to forward packets based on names. These new routers must work together with conventional IP routers and name servers, participating in both IP routing and routing based on names. This integration is the basis of the TRIAD content layer. The content routing proposal is based on mapping URLs on next hops. In fact, routing operates on the granularity of server names instead of complete URLs. Hence, routing is based on the longest suffix of FQDNs (Fully Qualified Domain Name) gateways (firewalls / NATs) between different areas and on BGP routers between autonomous systems (ASes).

The Name Based Routing Protocol (NBRP) is the proposed protocol to perform routing in FQDN names. It follows a similar structure of BGP distributing IP prefixes reachability information through different ASes, NBRP distribute information about accessibility to name suffixes between content routers. Like BGP, NBRP uses a distance vector routing algorithm with information about path through the routers content to reach a content server. The forwarding state is loaded on the intermediaries content routers along the path where IP packets would be routed. Scalability is achieved by means of aggregating structured content names. Nevertheless, these mechanisms may fail if the data location does not follow the DNS hierarchy. To overcome this problem, TRIAD proposes a resolution mechanism from names to locators.

4.4 DONA

The Date Oriented Network Architecture (DONA) [24] is a project of University of California, Berkeley, supported by the NSF and British Telecom. It explores an alternative content-based network architecture that allows a client to request for a data using its name (a self-certificate label based on the data itself), not the location where the data is stored. For this, the architecture exposes two fundamental operational primitives:

- FIND: allows a client requests a particular piece of data by its name;
- REGISTER: indicates the content provider's desires to offer a particular data object.

To enable these primitives, DONA introduces a new class of network entity named Data Handler (DH) that combines name resolution and data caching functionalities. DHs are responsible for forwarding the requests to the nearest node and make data copies. Data transport is done over IP.

DONA offers an anycast service with indirection and caching options and the biggest challenge is the scalability capacity to store and solve all the content's identifiers.

4.5 CCN

The Content-Centric Networking (CCN) project [25] was funded by DARPA and is led by Van Jacobson at Palo Alto Research Center (PARC). Van Jacobson was among the first visionaries who made the call to look at the Future Internet from a content point of view. Only recently, details about the proposed architecture were published by the PARCH research group, which present a pragmatic strategy for gradually adoption.

CCN highlights the following issues as the underlying conflicts between the IP model focused on host locators and a network used for content dissemination:

- Availability: The fast and reliable access to content requires specific mechanisms from applications and content providers similar to implemented on the CDNs (Content Deliver Networks) and P2P (Peer-to-Peer networks) imposing additional operational and bandwidth costs;
- Safety: Confidence in the authenticity and integrity of the information currently are based on inconsistent information and are easily changed as the location of the data or the entities of a connection;

- Dependence of location: The mappings of contents on the location where they are stored complicate the design and implementation of network services.

The CCN communication model is driven by consumers of data, i.e., information's receivers. CCN defines two types of packets: a packet data and an interest packet. The receiver/consumer asks for content sending a broadcast of messages of interest through all available interfaces. Any node that receives the messages and contain the requested data can answer with a data packet. This means that data packets are transmitted only in response to packet of interest and they are cleared of the router as the data is consumed. The packet forwarding follows a pattern similar to name-based routing proposed in TRIAD, where the content's reachability needs to be propagated by a routing protocol. A key difference is that by virtue of the built-in content support routing in CCN, it does not need to worry about loops or multiple paths through the network to a destination. One of the great challenges of the CCN model is that it requires that interest messages being temporarily stored at the intermediate nodes until being consumed, which require considerable (per-flow) state at routers. In return, CCN makes better use of temporal memory (buffers); the same packet is not stored multiple times and can be used to meet the requirements of various consumers in parallel or temporally separated.

4.6 PSIRP

The PSIRP (Publish/Subscribe Internet Routing Paradigm) is a project [26] being coordinated by HelsinkiUniversity of Technology (TKK) and Helsinki Institute for Information Technology (HIIT) with duration of 30 months (January 2008 – June 2010). It addresses the current weaknesses of the Internet by proposing a fundamental reform of its paradigms and enabling technologies. According to PSIRP, the main flaw in the Internet design is the imbalance in favor of the transmitter of information: the network accepts any packets sent by the transmitter and makes the best effort to deliver them to the receiver. This has led to increasing problems with spam mails and DoS attacks, forcing users to hide their email addresses and fragment network connectivity with firewalls.

PSIRP proposes a pure publish/subscribe paradigm as a solution for this problem. In this approach, the transmitters publish what they want to send and the receivers subscribe to the publications that they want to receive. Under this networking model, security, mobility and multicast are native and only the information needs to be named. The project explores a reformulation of TCP/IP layers based on this paradigm.

The architecture design is extensively validated by experimental research activities. Two different approaches are explored: (i) pub/sub running as an overlay layer to the IP (evolutionary approach), and (ii) IP layer being also completely replaced (clean-slate approach). Noteworthy, economic incentives for adoption including the market roles of different players are being considered in parallel to the design and implementation work, which brings a new dimension of great value.

The design principles adopted by the project can be summarized as follows [27]:

- PS1 - Information is hierarchically organized, starting with forwarding identifiers of limited meaning to higher level concepts (e.g. ontology).
- PS2 - Definition of information scoping on the different semantics levels using Rendezvous, discovery, research, and others functionalities.
- PS3 - The architecture is neutral to semantics and data structure, data is transmitted based only on the identifiers.
- PS4 - The architecture is receiver-oriented: the hosts only receive data, if they agreed in to receive it in advance via signaling protocol.

One design choice in PSIRP is RTFM [28], which takes its name from its functional blocks defined recursively. Rendezvous (R) functional block is in charge of matching the subscriptions to the publications as well as defining the information fields. Topology (T) functional block creates and maintains the trees connectivity for traffic routing. Forwarding (F) makes the forwarding of the data based on new identifiers of the multicast trees. Finally, Mediation and More (M) refer to other transmission functions, e.g. network coding or caching.

RTFM operation starts with a node sending a message to subscribe to a publication using an information identifier from the highest level. This message is distributed by the Rendezvous mechanism to find a copy of the metadata of the publication. In this process, the distributed topology management system gathers enough information to identify the trees necessaries to deliver data to the subscriber (s). RTF functional blocks are distributed and natively recursive.

4.7 Haggle

Haggle is a FP6 four-year project, started in 2006, to develop new network architecture design to enable communication between autonomic mobile devices with intermittent network connectivity, exploiting the paradigm of opportunistic communications [29]. The proposal eliminates the layers above the link layer and uses the applications' messages directly to forward information, eliminating this functionality from the network layer. The delivery of the messages is based on the best effort principle and use the context information to forward the messages between mobile devices with local/or intermittent connectivity. Haggle is based on application layer information, i.e. data-centric not host-centric.

4.8 Postcards from the Edge

NSF FIND Postcards from the Edge project aims to design a cache-and-forward network architecture based on computation and storage of heterogeneous systems [30]. The main objective is a transport service and a hop-by-hop opportunistic forwarding of large files. The experimental validation of the project will be use Planet-Lab, an academic testbed, as well ORBIT, a testbed with wireless technologies.

4.9 Scaffold

Scaffold (Service-Centric Architecture For Flexible Object Localization and Distribution) [31] is a new project supported by NSF and Cisco, started in 2010, towards a new network architecture based on direct addressing/naming (potentially distributed or replicated) from information objects or services. Instead of employing multi-layered ad-hoc techniques, Scaffold focuses on treat various types of churn, e.g. node failures, planned maintenance, load balancing, migration of workload and physical mobility, challenges typically arise in network services such cloud data centers. Although the architecture design has a "clean-slate" approach, the link with the current infrastructure and an incremental adoption is an immediate focus on the project, looking at the possibilities and limitations of data center architectures and enterprise networks as well as infrastructure networks with a single operator but geographically distributed.

5 Conclusion

A large number of research projects around the World are tackling the future Internet architectural issues from diverse perspectives. Given the grand scale of alternative approaches, a major effort by the research community itself is called for to start converging proposals into more pragmatic proposals. At this point, market and economical issues will play a fundamental role in determining whether and how any concrete features make their way up to planet-scale running code. To this end, experimental research at scale needs to proof the technical viability of the evolved Internet. Finally, if the right incentives for adoption are given, our next generation may seamlessly enjoy an improved Internet experience as a granted utility of the 21st century.

References

1. Clark, D., Sollins, K., Wroclawski, J., et al.: New Arch: Future Generation Internet Architecture DARPA Final Technical Report (2003),
   http://www.isi.edu/newarch/ (accessed April 29, 2010)
2. Dempsey, H.: GENI: Global Environment for Network Innovations. In: Future of the Internet Conference, Bled, Slovenia (2008)
3. Elliot, C.: GENI Project and New Network Architectures for the Future Internet (2009),
   http://www.cpqd.com.br/futurodainternet/programacao.html (accessed April 20, 2010)
4. The OpenFlow Switch Consortium, http://www.openflowswitch.org/ (accessed April 29, 2010)
5. Galis, A., Abramowicz, H., Brunner, M., et al.: Management and Service-aware Networking Architectures (MANA) for Future Internet: System Functions, Capabilities and Requirements. MANA Position Paper v6.0 (2009)

6. AKARI, New Generation Network Architecture AKARI Conceptual Design. Project Description v1.1 (2008), http://akari-project.nict.go.jp/eng/conceptdesign.htm (accessed April 20, 2010)
7. Harai, H.: New-Generation Network vision of NICT and the Akari project Workshop: New Architectures for Future Internet (2009), http://www.cpqd.com.br/futurodainternet/programacao.html (Accessed April 20, 2010)
8. Tronco, T.: ARCMIP Project Workshop: New Architectures for Future Internet (2009), http://www.cpqd.com.br/futurodainternet/programacao.html (accessed April 20, 2010)
9. Salvador, M.: Project GIGA – Optical Networks Experimental Research for the Future Internet International Workshop: New Architectures for Future Internet (2009), http://www.cpqd.com.br/futurodainternet/programacao.html (accessed April 20, 2010)
10. Horizon project, http://www.gta.ufrj.br/horizon/index.php/members (accessed April 20, 2010)
11. KyaTera Network, http://www.kyatera.fapesp.br/ (accessed April 20, 2010)
12. RNP, http://www.rnp.br/en/ (Accessed April 20, 2010)
13. Kazhamiakin, R., Bertoli, P., Paolucci, M., Pistore1, M., Wagner, M.: YourWay!: a Platform for Composing and Executing Services Driven by User Resources (2008), http://ftp.informatik.rwth-aachen.de/Publications/CEUR-WS/Vol-399/paper05.pdf (Accessed April 30, 2010)
14. Daidalos project, http://www.ist-daidalos.org/
15. Context Casting (C-CAST), http://www.ict-ccast.eu/
16. The Future of the Internet, A Compendium of European Projects on ICT Research Supported by the EU 7th Framework Programme for RTD (2008)
17. Deshpande, A., Nath, S., Gibbons, P., Seshan, S.: IRIS: Internet-scale Resource-Intensive Sensor Services. In: ACM SIGMOD 2003 (2003), http://www.intel-iris.net/papers/irisnet-demo-sigmod03.pdf
18. Shneidman, J., Pietzuch, P., Ledlie, J., Roussopoulos, M., Seltzer, M., Welsh, M.: Hourglass: An Infrastructure for Connecting Sensor Networks and Applications. Harvard Technical Report TR-21-04 (2004), http://www.eecs.harvard.edu/~syrah/hourglass/papers/tr2104.pdf (accessed April 30, 2010)
19. SENSEI, Deliverable report WP3/D3.1. State of the Art – Sensor Frameworks and Future Internet (2008)
20. Egan, R.: SENSEI: Sensor and ubiquitous connectivity. In: EU-Japan Simposium on Future Networks (2008)
21. Stoica, I., Adkins, D., Zhuang, S., Shenker, S., Surana, S.: Internet indirection infrastructure. In: SIGCOMM 2002, New York, NY, USA (2002)
22. Balakrishnan, H., Lakshminarayanan, K., Ratnasamy, S., Shenker, S., Stoica, I., Walsh, M.: A layered naming architecture for the internet. In: SIGCOMM 2004, New York, NY, USA (2004)
23. Gritter, M., Cheriton, D.: An architecture for content routing support in the Internet. In: USITS 2001: Proceedings of the 3rd conference on USENIX Symposium on Internet Technologies and Systems, Berkeley, CA,USA. USENIX Association (2001)

24. Koponen, T., Chawla, M., Chun, B., Ermolinskiy, A., Kim, K., Shenker, S., Stoica, I.: A data-oriented (and beyond) network architecture. SIGCOMM Comput. Commun. Rev. 37(4), 181–192 (2007)
25. Jacobson, V., Smetters, D., Thornton, J., Plass, M., Briggs, N., Braynard, R.: Networking Named Content. In: CoNEXT 2009, Rome, Italy (December 2009), http://www.parc.com/content/attachments/networking-named-content-preprint2.pdf (accessed in April 30, 2010)
26. Trossen, D. (ed.): Conceptual Architecture of PSIRP Including Subcomponent Descriptions (D2.2) (2008), http://psirp.org/publications
27. Tarkoma, S., Trossen, D., Särelä, M.: Black boxes: making ends meet in data driven networking. In: Proceedings of the 3rd international Workshop on Mobility in the Evolving internet Architecture, MobiArch 2008, Seattle, WA, USA, August 22, pp. 67–72. ACM, New York (2008)
28. Särelä, M., Rinta-aho, T., Tarkoma, T.: RTFM: Publish/subscribe internetworking architecture. In: ICT Mobile Summit, Stockholm (2008)
29. Nordström, E., Gunningberg, P., Rohner, C.: Haggle: a data-centric network architecture for mobile devices. In: Proceedings of the 2009 Mobihoc, New Orleans, Louisiana, USA (2009)
30. Paul, S.: Postcards from the edge: A cache-and-forward architecture for the future internet. In: NeXtworking 2007, 2nd COST-NSF Workshop on Future Internet, Berlin, Germany (2007)
31. Freedman, M.: Building a service-centric network with SCAFFOLD Princeton University (2010), http://netseminar.stanford.edu/seminars/freedman-scaffold.pdf (accessed April 30, 2010)

OneLab: An Open Federated Facility for Experimentally Driven Future Internet Research

Serge Fdida[1], Timur Friedman[1], and Thierry Parmentelat[2]

[1] UPMC Paris Universitas and the CNRS, Paris, France
sergefdida@lip6.fr, timur.friedman@upmc.fr
[2] INRIA, Sophia Antipolis, France
thierry.parmentelat@sophia.inria.fr

Abstract. Several initiatives worldwide are seeking to build an open, general-purpose, and sustainable large-scale shared experimental facility to foster the emergence of the Future Internet. This objective is ambitious as it calls for the setting up of testbeds to study solutions yet to be designed. Furthermore, any proposed new architecture must be accompanied by a transition scenario to overcome the significant obstacles that will lie in the path to its eventual adoption. The OneLab experimental facility is a leading prototype for a flexible federation of testbeds that is open to the current Internet. OneLab has pioneered the concept of testbed federation, providing a federation model that has been proven through a durable interconnection between its flagship testbed PlanetLab Europe (PLE) and the global PlanetLab infrastructure, mutualising over five hundred sites around the world. OneLab is further developing an understanding of what it means for autonomous organizations operating heterogeneous testbeds to federate their computation, storage, and network resources, including defining terminology, establishing universal design principles, and identifying candidate federation strategies.

1 Introduction

Demand is increasing among researchers and production system architects to combine compute, storage, and network resources from multiple sources (e.g., an organization's own resources, their partners' resources, commercial and academic clouds, programmable network substrates). This objective has emerged in the framework of network testbeds developed to conduct experiments, but the situation resembles what emerging networks faced at the dawn of the Internet. Federation is perceived as a means to increase the utility of a testbed by providing access to a larger set of heterogeneous resources, scaling to large systems, adding geographical diversity, helping to reach sustainability, and benefiting from best practices.

Many of today's testbeds address a given technology, an emerging service, a near-term product, or an important management issue, but suffer from a lack of sustainability, of support for longer-term research, and/or do not have international visibility. The Future Internet will be polymorphic, aggregating numerous types of systems. Yet there are currently few possibilities to experiment in a hybrid environment. A few projects have taken the first steps to demonstrate the potential effectiveness of such combinations, but substantial concerns remain about security, interoperability, tools, and management.

OneLab [32] addresses issues related to federating resources from multiple autonomous organizations into a global shared resource pool with a standardized interface to access them. OneLab already enjoys global scale through federation and it is running testbeds allowing experimentation with different technologies to meet the variety of needs of a broad customer base. It is developing a single access model to a diversity of networking technologies, it allows resources to be shared through the powerful paradigm of virtualization wherever this is possible, and it is extending its federation model to cover an array of heterogeneous testbeds, thereby lowering the entry cost to each individual facility. Achieving such federation is a challenge, as it requires solutions to issues of identity management, authentication and authorization, resource description, policy specification and enforcement, economics and incentives, virtualization technologies, operations and management, user-level abstractions and services, and governance considerations. Building the facility requires research on the architecture of the system as well as on the tools needed to operate it and provide accurate and secure data to its users.

Sec. 2 of this chapter reviews the context and history of OneLab while Sec. 3 presents the current status of OneLab's federation of testbeds, and Sec. 4 describes the larger ecosystem in which OneLab operates. OneLab's work on the federation approach is outlined in Sec. 5. Sec. 6 concludes the chapter.

2 Context and History

The rationale for OneLab and its federation approach have their origins in the E-NEXT project. In 2003, this project, a Network of Excellence (NoE) funded through the European Union's Sixth Framework Programme (FP6), defined a vision for building an experimental facility for Future Internet research. The chosen starting point was the existing highly successful PlanetLab testbed [38, 39, 40, 8], which was already deployed at a global scale with hundreds of users. A general-purpose experimental facility would then be built by gradually extending the testbed's capabilities and through federation, by integrating existing and new testbeds that support research on networks and services. The creator of PlanetLab, Princeton University professor Larry Peterson, lent his support to this vision in March 2004.

Work to turn this vision into reality began in September 2006 with the start of the FP6 STREP project eponymously entitled OneLab. Important aspects of this vision were also adopted by major programs that were launched in 2007. In Europe, the Seventh Framework Programme (FP7) set up the FIRE Initiative [17], which OneLab then joined. In the US, the National Science Foundation (NSF) inaugurated the GENI Initiative [18].

OneLab's 1.9 M€ STREP funding (2006-2008) allowed it to develop a proof of concept. Further funding of 6.3 M€ for OneLab2, an FP7 Integrated Project (2008-2010), has established OneLab as a leading prototype for the FIRE experimental facility. Through these two grants, OneLab created PlanetLab Europe [41] (PLE) as its flagship testbed. PLE is an autonomous European testbed, federated with PlanetLab Central in the US, and in the process of federating with a number of other PlanetLab-like testbeds worldwide. OneLab also defined a framework and built technology for extending PlanetLab into new environments (notably wireless and emulation environments), and built up the monitoring capabilities of the system, providing a set of tools for experimenters and testbed administrators that gives them vital information about testbed conditions.

The activities of FIRE, GENI, and other initiatives to develop prototype experimental facilities to support research on the future of the Internet reflect the importance of testbeds, in addition to network science, in providing experimentation services to the Future Internet research community at large. The goal for experimental federated testbeds is to enable large scale and diverse experiments with Future Internet technologies, from components to complete systems, and to validate and compare them with existing or evolving solutions.

A federation of testbeds aims at creating a physical and logical interconnection of several independent experimental facilities or testbeds to provide a larger-scale, more diverse and/or higher performance platform for carrying out advanced tests and experiments. Testbed federation is used to enrich the environment for testing and experimentation beyond what experimenters can access through individual independent testbeds. Federation is defined by the presence of at least some common objectives of the testbeds to be federated. Federation should support access to several platforms, networks, and services for testing in a broader context, e.g., for scalability, interoperability, or system-level testing. It enables trial and evaluation of service concepts, technologies, system solutions, and business models. Federation strategies can be described as horizontal (supporting large scale experiments with a diversity of end users) or vertical (supporting experiments across networking and service platform layers). The relationships between testbeds can also be seen as peer relationships or customer-provider relationships.

The federation of independent network experimental facilities is perhaps the only meaningful way to achieve the required scale, geographic coverage, and realism for supporting Future Internet research.

3 OneLab Today

This section describes the OneLab experimental facility as it exists today. Sec. 3.1 describes PlanetLab Europe (PLE), OneLab's flagship testbed. Sec. 3.2 describes the NITOS wireless testbed. Sec. 3.3 describes OneLab's federation framework, SFA. And Sec. 3.4 describes the research tools developed for OneLab.

3.1 PlanetLab Europe (PLE)

PlanetLab Europe [41], or PLE, is OneLab's flagship testbed. This is a platform for testing novel ideas in network overlays, content distribution systems, distributed systems, and peer-to-peer technology. As of this writing, it consists of over 160 server-class computers, or *nodes*, at around 80 sites across Europe, with a few beyond Europe, administered from the PLE **operations centre** at OneLab lead partner UPMC's premises in Paris. Each node runs the PlanetLab operating system, which is based upon the Linux-VServer virtualization layer. PlanetLab Europe is a slice-based facility: a researcher obtains a slice across the system that consists of virtual machines on any or all of the nodes. The researcher can log in to each of these virtual machines via the SSH secure remote login tool, and find himself as the root user in a Fedora 8 Linux environment. He can deploy whichever software he likes on these virtual machines, subject only to a few restrictions based on the shared kernel.

Since the nodes are open to the public Internet, researchers can experiment with distributed applications in a real-life testing environment. Furthermore, researchers can deploy services that are used by regular Internet end-users worldwide. For example, one content distribution experiment on PlanetLab offers faster web downloading to thousands of end-users in countries across the world. The researchers who deployed this service use it to study application performance and end-user behaviour.

A key benefit of PLE to researchers is that it allows them to deploy their experiments at a global scale, exposing their applications to geographic and network topological diversity and allowing them to deploy services in proximity to end-users. While the testbed's own nodes are scattered essentially across Europe, federation with the global PlanetLab system [40] provides European researchers with access to the combined system, which as of this writing consists of more than 1,000 nodes at over 500 sites worldwide.

In the near term, PLE is scheduled to incorporate EverLab [15], a PlanetLab-based system that was developed as part of the FP6 EVERGROW [14] Integrated Project. EverLab differs from a typical PlanetLab system in that each site provides a computing cluster of a dozen or more servers, rather than simply two servers. EverLab consists of clusters at several sites in Europe and Israel.

The code underlying PlanetLab was created at Princeton University. Since 2006, OneLab partner INRIA has become an equal partner with Princeton in developing that code. The **OneLab Build** of the PlanetLab software [31] includes extensions that are specific to the OneLab facility. Among these, OneLab has contributed the dummynet [6] module for Linux that allows an experimenter to emulate specified loss and delay profiles for traffic entering and leaving the virtual machines on their slice. INRIA-Princeton co-development is the rule, however, and the software is hosted under a common codebase that serves both FIRE, through the OneLab facility, and GENI, where it is one of the five control frameworks around which that initiative is clustered.

For a testbed to be open to the public Internet, involve real-world end-users, and enter into federation agreements with other testbeds, it requires institutional backing. In August 2008, core OneLab partners UPMC and INRIA established the **PlanetLab Europe Consortium** as a partnership responsible for running the testbed. This partnership has no defined end-date, so it can assume long-term supervision of the testbed in a way that individual grant-based projects (such as the OneLab STREP and the OneLab2 Integrated Project) cannot. The Consortium is currently being expanded to include OneLab partners Hebrew University of Jerusalem (HUJI) and the University of Pisa. A memorandum of understanding guides the joint federation work of the PLE Consortium and the global PlanetLab Consortium.

Over 45 institutions in Europe, representing nearly 400 researchers, have signed user membership contracts with the PLE Consortium. To join as a user member, an institution contributes two server-class computers that become PLE nodes. It makes these nodes available from their networks to be freely accessed by researchers via the Internet. Because this openness requires responsible behaviour, member institutions legally commit their researchers to follow an acceptable use policy.

To guide researchers on PLE, HUJI has developed a set of **user tutorials**. These include videos that are available through the PLE online channels:

- Dailymotion http://www.dailymotion.com/PlanetLabEurope
- YouTube http://www.youtube.com/user/PlanetLabEurope

3.2 The NITOS Testbed

The OneLab facility also includes a wireless testbed that is freely available to researchers. The **NITOS** testbed [28] or Network Implementation Testbed using Open-Source code, run by OneLab partner CERTH, consists of 50 nodes that are deployed on the campus of the University of Thessaly, located both inside and outside the university's NITLab building. These nodes primarily provide 802.11 connectivity. (OneLab includes two other wireless testbeds, run by ETH Zurich and INRIA. These employ the same management framework as NITOS, and are used internally for proof of new concepts.)

The OneLab wireless testbeds are managed by the **OMF** control, management, and measurement framework [29]. OMF was originally developed for the ORBIT testbed [33] at Rutgers University in the US. Since then, OneLab partner NICTA has taken over its development, advancing it both in FIRE, in the context of OneLab, and in GENI, where it is one of the five control frameworks around which the initiative is clustered. Under NICTA's direction, OMF has been deployed in over 15 testbeds worldwide. In these testbeds, it controls networking resources as diverse as WiMAX, software defined radios, sensor networks, and various access networking technologies. OMF provides a holistic solution supporting the experimenter and the testbed operator alike as well as providing a fully integrated measurement framework.

3.3 The SFA Federation Framework

OneLab federates testbeds using an emerging standard framework called SFA [44], for Slice-based Facility Architecture. PlanetLab developed an SFA, and OMF followed suit. The implementations share a common high-level interface, but differ in many specifics, such as how resources are described. OneLab is working on harmonizing these details in a technically sound manner.

Authentication-based authorization of testbed users is at the heart of federation under SFA. Two testbeds, A and B, are federated if the users of A are able to access a slice (a set of resources) in B, and vice versa for symmetric federation. Authentication is performed through X-509 certificates. Authorization is performed in a distributed manner, based on policies [4] that are defined locally by the resource owner, based on information about the requester. The key architectural features of SFA therefore consist in the authentication mechanism and in the language for describing resources, requesting them, and granting slice access.

3.4 Research Tools

There are almost as many experiment control systems as there are testbeds today. Any testbed of sufficient size and complexity requires a system to facilitate the following tasks:

- Setting up of experiments.
- Control of experiments while they are running.
- Retrieval, for analysis, of the data and meta-data associated with an experiment after it has completed.

These three tasks are grounded in a fourth task: measurement. When research is conducted in an uncontrolled environment (such as in PLE and NITOS), measurement is key to helping a researcher understand the impact of network conditions on his experiments. We describe as *research tools* the set of tools that help accomplish the tasks listed above. OneLab provides research tools on both the PLE and NITOS testbeds.

PLE research tools PLE offers the following research tools:

- **MySlice** [26]. MySlice, developed by UPMC, gives users both a web interface and a programmable API through which to manage their slices: starting with slice setup, through experiment run-time, to retrospective analysis of experimental data after an experiment is concluded. MySlice will also offer visualization tools for collected data. Network-related data for MySlice comes from TopHat, described below, and PlanetLab-node-related data comes from Princeton's Co-Mon [6] service.
- **TopHat** [5, 46]. Because the PlanetLab platform consists of end-systems, users do not have a privileged view of the network that lies between the nodes. TopHat, run by UPMC, is a service that provides information on the network topology, including characteristics of the topology such as latencies and available bandwidth. It draws upon its own dedicated set of agents, deployed in a slice on

PlanetLab, as well as on the ETOMIC, SONoMA, and DIMES infrastructures described below, and other external data sources.

- **ETOMIC [27, 13] and SONoMA [45].** These infrastructures consist of specialized high precision measurement boxes deployed at a few dozen sites across Europe, including at a number of PLE sites. They have been developed by OneLab partners Eötvös Loránd University (ELTE) and Universidad Autónoma de Madrid, initially as part of the FP6 EVERGROW[14] project. The boxes' clocks are synchronized by GPS. ETOMIC boxes, which must be reserved to be used, have customized FPGA cards that can obtain measurements with a resolution of tens of nanoseconds. SONoMA boxes do not require such cards and can be used on demand, without reservations, and offer a resolution of tens of microseconds. The GPS synchronization and precision of these systems allow for improved available bandwidth and one way delay estimations.
- **DIMES [10].** Experimental applications on PLE interact with hosts across the Internet. Experimenters' need for topological information therefore extends to the Internet as a whole. The DIMES infrastructure, run by OneLab partner Tel-Aviv University (TAU), consists of thousands of agents deployed around the world, providing a more comprehensive picture than can be provided by agents deployed on PlanetLab alone.
- **PLE packet tracking [36].** This is a passive monitoring infrastructure that is deployed in the ETOMIC boxes and that can be deployed in other equipment as well. It is developed by OneLab partners Quantavis, Fraunhofer Fokus, and CINI. The packet tracking system is based on CoMo [9] open-source software (not to be confused with CoMon, mentioned previously), which offers researchers the ability to deploy their own scripts that determine which of their own packet information they retrieve, while at the same time allowing network operators who host the boxes to maintain the privacy of other users' packets.
- **Network measurements virtual observatory.** This tool is expected to be available in a few months from ELTE. It indexes data that has been collected by the measurement infrastructures described above, as well as by other infrastructures. The virtual observatory thereby allows researchers to conduct retrospective analyses.
- **Visualization tools.** TAU will be providing tools to aid in retrospective data analysis through visualization.

NITOS research tools The NITOS testbed is based on the OMF framework, as described earlier. OMF integrates seamlessly with the **OML** measurement framework [30]. OML is used to collect the full range of measurements associated with a NITOS experiment, including packet traces, wireless signal strengths, and node mobility, as well as instrumented user applications. It also provides support for disconnected operation, which is especially important for mobility-related experiments. As OML can collect measurements across the entire software stack it has found applications outside the testbed domain, such as for the continuous monitoring of distributed services.

4 The OneLab Ecosystem

Sec. 2 mentioned the important and growing effort to design, operate, and extend network testbeds at the international level. A number of initiatives are being developed worldwide at the national or regional level. FIRE and GENI illustrate this, as do, notably, AKARI [2] in Japan, G-Lab [20] in Germany, and the Pan-Asian AsiaFI [3]. Within FIRE, several testbed projects are currently being deployed. In addition to OneLab, these currently include Panlab [37], FEDERICA [16], WISEBED [48], and VITAL++ [47].

OneLab has distinguished itself for its role in pioneering the federation concept. Among the European projects, to date only OneLab has developed a working federation. (In the US, GENI has been pursuing this goal as well, advancing a set of five control frameworks, each with a federation architecture [19].) The federation of wired and wireless networks has been part of the OneLab vision since the beginning, and OneLab is now set to push federation towards a diverse set of systems, with the goal of providing access to a wide variety of heterogeneous resources through a common interface. Current plans involve the federation of PlanetLab-based and other systems around PLE and the federation of OMF-based systems around NITOS.

Around PlanetLab Europe. OneLab currently mutualizes resources from **PlanetLab Europe** and the PlanetLab Central. Extension of the OneLab federation starts with additional PlanetLab-based testbeds:

- **PlanetLab Japan** [42], or PLJ, run by OneLab partner NICT. PLJ is associated with Japan's JGN2plus advanced network [24].
- **PPK** [42], or Private PlanetLab Korea, run by KAIST. PPK provides testbed services for users of Korea's KOREN advanced network [25].
- **6P-UOA** [1] run by OneLab partner Tsinghua University, in association with the China Education and Research Network (CERNET) [7]. This testbed uses PlanetLab technology at the control centre, but uses a time-sharing rather than a virtualization architecture for its nodes, which operate in an IPv6 environment.
- **G-Lab** [20], the large Future Internet project funded by the German Federal Ministry of Education and Research.
- **EmanicsLab** [11], a PlanetLab system that was developed for the EMANICS FP6 NoE on management of the Future Internet.

In addition, OneLab is actively pursuing PLE-FEDERICA federation. FEDERICA [16] is an FP7 testbed run by a group of European NRENs. It consists of several sites in Europe with reserved optical links between them. FEDERICA allows users to deploy experimental routing protocols on programmable routers.

Around NITOS. Through OMF, NITOS is capable of federation with other OMF-based testbeds, and notably **ORBIT** [33], a wireless network emulator based at Rutgers WINLAB in the United States, consisting of a grid of 400 802.11 radio nodes that can be dynamically interconnected into specified topologies with reproducible wireless channel models. This testbed stands out for its large number of devices and for its highly controlled and configurable environment. Other OMF-based testbeds are also scheduled to federate within OneLab.

5 Work in Progress on OneLab Federation

OneLab is well rooted in the current efforts to build a Future Internet facility. Historically, platforms for research in networking [40, 41, 33, 12, 23, 22] and to a lesser extent in distributed computing [34, 21] have more or less followed a similar path: the first step is to focus on the deployment (make it work), then effort is placed on repackaging software so that others can set up their own local or consortium-wide instance (make it reusable); as soon as several instances of the same software exist round the world, it is tempting to connect them in order to optimize operations and hardware cost. This sort of limited federation has already been achieved in various circumstances, with PlanetLab Europe and PlanetLab Central pioneering this approach in 2007. The current challenge is now much more ambitious, as the need for smoothly integrating various kinds of resources arises.

In our vision, a complete federation scheme has the following advantages: (a) each user only needs a single point of entry into the system (single sign-on); (b) we advocate that this entry point does not need to be unique, i.e., each testbed in the system can act as the entry point for its "natural" users; (c) this way, users and resource providers need to only sign a legal contract with one testbed, which in turn has agreements with other testbeds, thus forming by transitivity the legal links between all the players in the system; (d) each testbed remains free to decide on its own policies, which can leverage peering agreements among testbeds; (e) from a single user interface, each user gets a clear view of the resources he might be granted an access to, and can provision and control them in accordance with local policies; (f) this resource-sharing groundwork will allow for the construction of common higher-level user tools, such as experiment control tools, that help in deploying and running experiments.

There are many challenges to be met in creating a global federation. Among them are:

- **Economics:** There clearly is a need to get a better understanding of what users really need, of what they are ready to give for obtaining that, and of how an economic model can help in creating incentives for people to attach resources in such a federation. The PlanetLab experience has shed some light on this matter, but clearly much more heterogeneous resources, as well as a much larger federation, shows the need for further investigation in this area.
- **Policies:** In a similar fashion, we need to be sure that every resource owner has the right set of tools to decide on its own policy for sharing its resources; so far we have been successful in implementing reasonably simple policies, but more work is needed to address a wider variety of users concerns.
- **Scalability:** From the very beginning, building federation has tried to solve the n-square problem that quickly enters the picture as soon as a global architecture is defined. SFA addresses this particular issue through hierarchical

name spaces, not unlike DNS. At this point, subdomains still remain to be put into practice.

- **Migration and software reusability:** SFA has been implemented once from scratch in the PlanetLab case, and is being re-implemented in the OMF case; for that latter work, we have tried to reuse the former code as much as possible. We aim to gather expertise over time as to the most effective ways to make a pre-existing testbed SFA-compliant.
- **Technology spectrum:** We wish to bring as many different technologies to users as possible; at this point, we are targeting sensor networks, delay tolerant networks (DTNs), as well as more traditional technologies like cellular (3G) networks and optical networks at the European scale (i.e., being able to set up level 2 circuits over the whole federation), as well as emulation platforms.
- **User interface:** SFA at this point only offers users a command-line interface (CLI), and a graphical user interface (GUI) is lacking. One of the challenges here is that each testbed comes with its own way of presenting resource information, and users need to be assisted in managing this complexity beyond providing them with generic, but simplistic, XML-based tools. So we foresee that a plugin mechanism might become necessary to manage this type of diversity.
- **Research tools:** Once the basic, provisioning level of federation is established, relevant research tools that can operate across testbed boundaries will be a requirement. It is our belief that this domain will be subject to many cross-testbed discussions and cross-fertilization. As a first step, we are working on using OMF's tools to manage PlanetLab slices. Improvements in the research tools category are only just beginning.

6 Conclusion

Building a federated ecosystem of testbeds is a complex and risky process, with research, technological, economical and legal issues to address. The OneLab experimental facility is embracing this challenge together with a larger worldwide community and many fruitful and enthusiastic partnerships. The advantage of OneLab is that it is up and running and goes much beyond the operation of a vanilla PlanetLab, aiming to offer a federation of heterogeneous testbeds, wireless and wired.

Acknowledgments

We thank our OneLab colleagues for all of their contributions to this work.

The research leading to these results has received funding from the European Community's Seventh Framework Programme (FP7/2007-2013) under grant agreement n°224263-OneLab2.

References

1. 6P-UOA testbed, http://202.38.118.4/ (accessed March 2010)
2. AKARI initiative, Architecture Design Project for New Generation Network, Japan, http://akari-project.nict.go.jp/eng/index2.htm

3. AsiaFI, Asia Future Internet Forum, http://www.asiafi.net/
4. Bathia, S., Bavier, A., Peterson, L., Sevinc, S.: Establishing Resource Allocation Policies in Federated Systems,
   http://old.nabble.com/attachment/26322882/0/paper.pdf (accessed March 2010)
5. Bourgeau, T., Augé, J., Friedman, T.: TopHat: supporting experiments through measurement infrastructure federation. In: Proc. TridentCom (2010)
6. Carbone, M., Rizzo, L.: Dummynet revisited,
   http://info.iet.unipi.it/~luigi/papers/
   20091201-dummynet.pdf (accessed March 2010)
7. CERNET, China Education and Research Network,
   http://www.edu.cn/english/
8. Chun, B., Culler, T., Roscoe, A., Bavier, L., Peterson, M., Bowman, M.: PlanetLab: an overlay testbed for broad-coverage services. ACM SIGCOMM Computer Comm. Rev. 33(3), 3–12 (2003)
9. CoMo measurement platform, http://como.sourceforge.net/ (accessed March 2010)
10. DIMES measurement platform, http://www.netdimes.org/ (accessed March 2010)
11. EmanicsLab testbed, https://emanicslab.csg.uzh.ch/ (accessed March 2010)
12. Emulab network emulation testbed, http://www.emulab.net/ (accessed March 2010)
13. ETOMIC, European Traffic Observatory Measurement Infrastructure, http://www.etomic.org/ (accessed March 2010)
14. EVERGROW FP6 Integrated Project, http://www.evergrow.org/ (accessed March 2010)
15. EverLab testbed, http://www.everlab.org/ (accessed March 2010)
16. FEDERICA, Federated E-infrastructure Dedicated to European Researchers Innovating in Computing network Architectures, FP7 project,
    http://www.fp7-federica.eu/ (accessed March 2010)
17. The FIRE Initiative, of the European Seventh Framework Programme (FP7), http://cordis.europa.eu/fp7/ict/fire/ (accessed March 2010)
18. The GENI Initiative, of the US National Science Foundation (NSF), http://www.geni.net/ (accessed March 2010)
19. GENI Spiral One, http://groups.geni.net/geni/wiki/SpiralOne/ (accessed March 2010)
20. G-Lab testbed, Germany, http://www.german-lab.de/ (accessed March 2010)
21. GRID5000 testbed,
    https://www.grid5000.fr/mediawiki/index.php/Grid5000:Home (accessed March 2010)
22. HEN, Heterogeneous Experimental Network testbed,
    http://mediatools.cs.ucl.ac.uk/nets/hen (accessed March 2010)
23. IBBT BroadBand Communication Networks,
    http://www.ibbt.be/en/onderzoeksgroep/ugent-ibcn (accessed March 2010)
24. JGN2plus, Advanced Testbed Network for R&D,
    http://www.jgn.nict.go.jp/english/

25. KOREN, Korea Advanced Research Network,
    http://www.koren.kr/koren/eng/
26. MySlice prototype, http://myslice.planet-lab.eu/
27. Morato, D., Magana, E., Izal, M., Aracil, J., Naranjo, F., Astiz, F., Alonso, U., Csabai, I., Haga, P., Simon, G., Steger, J., Vattay, G.: The European Traffic Observatory Measurement Infrastructure (ETOMIC): A testbed for universal active and passive measurements. In: Proc. Tridentcom (2005)
28. NITOS, wireless testbed at the University of Thessaly,
    http://nitlab.inf.uth.gr/NIT/WirelessTestbed (accessed March 2010)
29. OMF, cOntrol and Management Framework, http://omf.mytestbed.net/
30. OML, OMF Measurement Library,
    http://omf.mytestbed.net/wiki/omf/Collecting_Measurements/ (accessed March 2010)
31. The OneLab code repository, http://build.onelab.eu/ (accessed March 2010)
32. The OneLab experimental facility, http://www.onelab.eu/ (accessed March 2010)
33. ORBIT, wireless testbed at Rutgers University, http://www.orbit-lab.org/ (accessed March 2010)
34. ORCA, Open Resource Control Architecture,
    http://nicl.cod.cs.duke.edu/orca/about.html (accessed March 2010)
35. Park, K., Pai, V.: CoMon: a mostly-scalable monitoring system for PlanetLab. SIGOPS Oper. Syst. Rev. 40(1), 65–74 (2006)
36. PlanetLab Europe Packet-tracking, http://wiki.packet-tracking.com/
37. Panlab, Pan European Laboratory Infrastructure Implementation,
    http://www.panlab.net/ (accessed March 2010)
38. Peterson, L., Bavier, A., Fiuczynski, M., Muir, S.: Experiences building PlanetLab. In: Proceedings OSDI (2006)
39. Peterson, L., Pai, V.: Experience-driven experimental systems research. Comm. of the ACM 50(11), 38–44 (2007)
40. PlanetLab Central testbed, http://www.planet-lab.org/ (accessed March 2010)
41. PlanetLab Europe testbed, http://www.planet-lab.eu/ (accessed March 2010)
42. PlanetLab Japan testbed, http://www.planet-lab.jp/ (accessed March 2010)
43. PPK, Private PlanetLab Korea testbed, http://www.planet-lab.kr/ (accessed March 2010)
44. SFA, Slice-Based Facility Architecture,
    http://svn.planet-lab.org/svn/sfa/trunk/docs/sfa.pdf (accessed March 2010)
45. SONoMA measurement infrastructure,
    http://www.complex.elte.hu/sonoma/ (accessed March 2010)
46. TopHat, a topology measurement system for testbed applications,
    http://top-hat.info/ (accessed March 2010)
47. VITAL++. Embedding P2P Technology in Next Generation Networks, FP7 project, http://www.ict-vitalpp.upatras.gr/ (accessed March 2010)
48. WISEBED, Wireless Sensor Network Testbeds, FP7 project,
    http://www.wisebed.eu/ (accessed March 2010)

RNP Experiences and Expectations in Future Internet Research and Development

Michael Stanton

Rede Nacional de Ensino e Pesquisa – RNP, Rua Lauro Muller 116/1103,
Botafogo 22290-906 Rio de Janeiro, RJ, Brazil
michael@rnp.br
(on secondment from Computing Institute, Universidade Federal Fluminense – UFF)

Abstract. RNP is Brazil's NREN (National Research and Education Network), fully supported by the federal government to provide advanced network services to the higher education and research community. RNP has operated its own IP network since 1992, and has continually renewed its technology since then. This chapter reports on the present state of RNP production infrastructure, including expectations for 2010. Additionally, a number of different Brazilian network testbed initiatives are presented, as well activities now being directed to Future Internet research and development.

1 Introduction to RNP

Electronic communication between computers reached Brazil in 1988, with the establishment of two international links to BITNET, and their extension to about 40 institutions by 1991 (Stanton 1993). RNP was created in 1989 as a project of what is now the Brazilian Ministry of Science and Technology (MCT), to deploy a national computer network connecting universities and research centres, and provide them with access to similar networks in other countries. The first version of the RNP national backbone network was deployed in 1992 using Internet (TCP/IP) technology, connecting points of presence (PoPs) in 11 capital cities – Brasil has 26 states and a Federal District (DF) containing the national capital – and providing an international connection to the USA (Stanton 1993).

During the 1990s, RNP continued renewing and extending its network until it reached all 27 capitals. By 1999, RNP's situation had altered. Instead of being a project of MCT, with consequent instability and insecurity for RNP staff and objectives, a non-profit private company, AsRNP, was constituted by the project participants, and was subsequently contracted by MCT to manage the national network. By 2002, AsRNP had been formally recognised as a "Social Organisation" by MCT, which legally permitted the ministry to sign long-term contracts without a tender process, and to administer its relations with AsRNP in a similar way to other specialised service-providing institutions (in scientific computing, astrophysics, synchrotron light, and so on) which fell in the general category of national laboratories. Of comparable importance was the cofinancing of AsRNP activities by the Ministry of Education (MEC), which provided by far the largest contingent of clients of the national network. In addition, the network had been restructured, using the recently introduced ATM and Frame Relay technologies, which permitted incremental adjustment of available bandwidth. This new version of the network, introduced in 2000, was known as RNP2, even though it was much more limited than the Internet2 network in the US. However, it represented a significant improvement, and was what was economically feasible at that time.

The stability brought about by this reorganisation has enabled RNP to determine and follow long-term objectives, and to continue greatly to expand its network and its activities, so that it can now be considered to be a world-class research network. Much of this has come about by the continued application of new, mainly optical, technologies to providing network connectivity, by developing and deploying advanced user services, and by taking advantages of economies of scale.

2 RNP's Current Connectivity

The current connectivity offered by RNP to its over 300 client institutions can be described under the following categories:

- Backbone network with one PoP per capital
- Direct connections to local PoP from institutions located outside capitals
- Community-based optical metro networks in capital cities
- International connectivity

2.1 Backbone Network with One PoP per Capital

Ever since the first version of the RNP backbone, the design has included a single PoP in each capital, usually a federal university which distributes connectivity within the local state, or the Federal District (DF). The first version on the network used 9.6 and 64 kbps links. The current version, which is the fifth and was commissioned in 2005, uses a variety of connections, ranging from 6 Mbps to 10 Gbps (see Fig. 1).

It can be seen that the network is built around a 10 city core, including links of 2.5 and 10 Gbps, implemented as non-redundant lambdas. The remaining 17 PoPs are linked to this core by leased point-to-point circuits, varying between 6 Mbps and 622 Mbps. Connections to the two northern capitals of Boa Vista and Macapá are currently made by satellite. The remaining links are all terrestrial.

Local connections to each PoP are managed by its own staff. The institution housing the PoP has an agreement with RNP for carrying out certain operational matters on RNP's behalf.

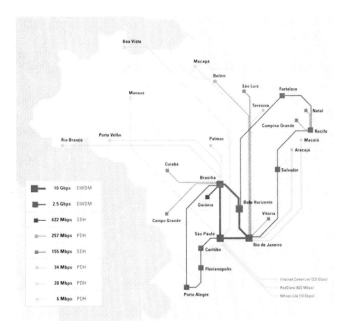

Fig. 1 Current version (v.5) of the RNP backbone network (28/11/2009) (Courtesy: RNP)

2.2 Direct Connections to Local PoP from Institutions Located Outside Capitals

Within each state (or the Federal District), connections between client institutions and the local PoP are handled in several ways. In some states, connectivity is provided by a statewide network, and is not an RNP responsibility. Additionally many institutions have multiple campi, of which only the main campus has traditionally been connected by RNP, it remaining the responsibility of the institution to solve its own internal connectivity problems.

In recent years, RNP has been charged with providing connectivity to the local PoP of an increasing number of federal institutions maintained by MCT and MEC, and located in non-capital cities. Most of these links are of low bandwidth (less than 4 Mbps), with the exception of federal universities and research units of MCT, which may have up to 155 Mbps connections.

2.3 Community-Based Optical Metro Networks in Capital Cities

RNP has been engaged in deploying its own optical fibre metro networks, initially in capital cities. This initiative was inspired by the example of the Canadian network, CANARIE, whose former chief architect, Bill St Arnaud, is a tireless proponent of community networks, where a collection of interested organisations pool their resources to build and run their own optical network. The business case for this was easy to make in Brazilian cities, due to the relatively high cost of urban links offered by telcos. It has been shown that the cost of investment could be recovered in 2 or 3 years from the savings on running costs, and the resulting capacities (minimum of 1 Gbps) were often thousands of times greater than the circuits which were being replaced.

The first such project was begun in 2004 in Belém, capital of the state of Pará. A feasibility study was carried out, and R$1.15M in funding to build the network was promised by MCT, and made available in December 2004 (Stanton 2005). Construction began early in 2005 and the optical infrastructure was completed in mid-2006, although the network, known as MetroBel, was only inaugurated in May, 2007, after the necessary network equipment had finally been imported. The resulting network linked 32 access points belonging to 12 separate institutions, each of which had dedicated access to a pair of fibres in the entire network covering 40 km. This enabled independent access to the PoP by each institution, and the possibility of building its own inter-campus network using the common infrastructure. In many cases, the network capacity to a campus was vastly increased: the 8 separate campi of the Universidade Federal do Pará, formerly linked at 128 kbps, migrated to 1 Gbps links, a factor of 8000 times the former capacity. A map of the MetroBel network is shown below (Fig. 2).

The MetroBel project has become a model for the rest of Brazil, and before the end of 2004 MCT had asked RNP to apply this model to all other 25 state capitals and Brasília, and made a further R$ 40M in funding available to carry this out. This much larger project is known as Redecomep, and a further 15 capitals have already inaugurated their networks by the end of 2009. The remaining 11 are expected to do so in 2010. RNP has also been asked to build similar networks in non-capital cities throughout the country, beginning with those housing more than one federal institution maintained by MCT or MEC.

Starting in 2001, RNP adhered to the current division of international Internet access between *collaboration*, with traffic between research and education (R&E) institutions, and *commodity*, which includes all other traffic. The global R&E Internet is accessed by dedicated international links between participating networks. RNP uses two such connections: one to the RedCLARA regional network, and a second to US networks.

The RedCLARA network was set up in 2004, largely financed by the EU through the ALICE project (Stöver 2003). ALICE continued until 2008, establishing an interconnection of 12 national R&E networks in Latin America, with access to GEANT in Europe. The complementary WHREN-LILA project, partially funded by the International Research Network Connections (IRNC) programme of the US government agency, NSF, linked RedCLARA to US networks on the East and West coasts, starting in 2005 (WHREN-LILA 2010).

The ALICE2 project, also financed by the EU, has led to the expansion of the RedCLARA network in 2009, with the inclusion of a 13th country. In Fig. 3, it can be seen that the main regional backbone is a 622 Mbps/1 Gbps ring interconnecting São Paulo (Brazil), Santiago (Chile), Panama City (Panama) and Miami (USA), with a 622 Mbps transatlantic connection to Madrid (Spain). Most national links to this network are 155 Mbps, although significant future changes will be discussed below.

RNP Experiences and Expectations in Future Internet Research and Development 157

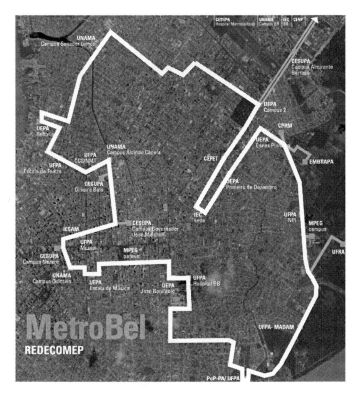

Fig. 2 The MetroBel network in Belém. The RNP PoP is located at the principal campus of UFPA, on the right side of the lower margin of the figure. (Courtesy: RNP)

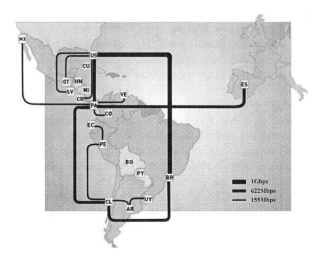

Fig. 3 RedCLARA topology in December 2009. The link shown between Brazil and the US is a subchannel of the Brazilian links to the US (see below). (Courtesy: CLARA)

A more direct impact on Brazil has been the result of altering the way RNP has dealt with international *collaboration* and *commodity* traffic to the US. By late 2008, RNP was purchasing 2 Gbps of international commodity transit in Brazil, from a large international provider, and this was insufficient. However, the price being paid was certainly sufficient to lease a 10 Gbps link between Brazil and the US, where the cost of commodity transit was not more than 25% of what was being paid in Brazil. It was decided to adopt this latter model, in which RNP was assisted by its US colleagues at the AMPATH exchange point in Miami. It should be noted that the ANSP network in São Paulo had reached a similar decision, and both networks reorganised their international links in 2009.

RNP and ANSP had been cooperating since the beginning of the NSF-supported WHREN-LILA project in 2005, of which both networks were partners in the shared use, together with RedCLARA, of a 2.5 Gbps link between São Paulo and Miami (WHREN-LILA 2009). This cooperation became closer over the years, and has been marked by collaboration in hosting the GLIF Open Lightpath Exchange (GOLE), called SouthernLight, used for international circuit management since 2008 (see later, below). The new agreement reached over the new international links was to pool (fully share) the two 10 Gbps links that the two organisations were to install between São Paulo and Miami. In addition, commodity transit in Miami would be bought from the same provider. As the two networks together already needed 3.5 Gbps of commodity transit, this was contracted in Miami for less than 10% of the price previously paid in São Paulo. The new arrangement has led to a more scalable access to commodity transit, and has effectively increased international collaboration bandwidth from 2.5 Gbps to about 15 Gbps, thus greatly increasing support for scientific collaboration.

3 New Network Infrastructure in 2010

2010 has been seen as the year that many of RNP infrastructure projects defined during the last few years would come to fruition. These included, naturally enough, the 27 optical metro networks in capital cities, expected to be operational in 2010. The metro networks were designed to complement a high-speed backbone network, also operating in Gbps. The Ipê network, as inaugurated in 2005, was seen within RNP as a major step forward, in bringing Gbps networking to 10 capital cities. However it was recognised that the next expansion would have to extend the Gbps core to many more capitals, and in 2008 a map of the expected future Ipê network in 2010 was devised, with 10 Gbps links to 18 capitals, and 1 Gbps links to the remaining 9. However, it was unclear as recently as early 2009 how this was to be accomplished.

This situation has been completely altered by the following recent development. In 2008, a takeover was mutually agreed of one Brazilian telco, Brasil Telecom, by another, Oi. However, the existing rules designed to promote competition in this sector did not allow such a combination. In spite of this, the regulatory authority, Anatel, gave its permission for the takeover, under several conditions, which included assisting RNP in its mission to connect its R&E clients for 10 years. RNP began discussing with Oi in early 2009 how this assistance could best be given, and, by early 2010, agreement had been reached for this to take the form of the provision of thirty-one optical circuits between pairs of capital cities. Of these, 11 would provide 3 GigE connections, and the remaining 20 would be fully transparent 10 Gbps lambdas. All capitals south of the River Amazon were to be included (see Fig. 4). The remaining 3 capitals will have to wait for the arrival in 2012 of electricity transmission lines from the south side of the river.

The other novelty expected in RNP's network connectivity in 2010 will be due to the CLARA initiative to seek sustainable network infrastructure by establishing long-term partnerships with the owners of optical fibre assets, prepared to share them with research networks, which would then purchase the optical networking equipment which would support such joint use.

The first two such cases will provide 10 Gbps links between Chile, Argentina and Brazil in 2010. Between Argentina and Chile, a small Argentine telco agreed to cooperate with CLARA, RNP and its Argentine equivalent, InnovaRed, using a fibre pair between Santiago and Buenos Aires. The optical transmission equipment installed, which was supplied by the Brazilian company, Padtec, will make it

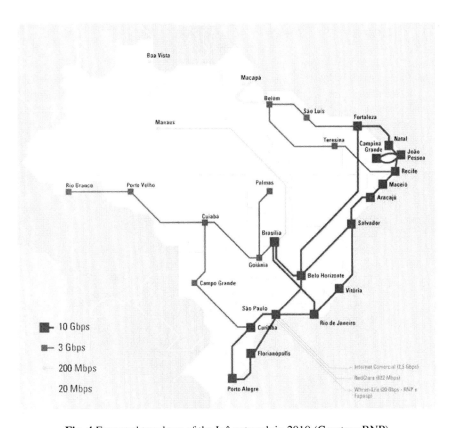

Fig. 4 Expected topology of the Ipê network in 2010 (Courtesy RNP)

possible to provide five 10 Gbps lambdas (optical circuits) along the entire route. A similar deal between RNP, InnovaRed and a large international telco was reached, for a fibre pair between Buenos Aires and Porto Alegre. In this case up to 8 lambdas are to be shared. In both cases, the CLARA network will gain a 10 Gbps link, and the local network (RNP or InnovaRed) will gain a further 10 Gbps for its own use, within its own country. Thus high-capacity international networking will be able to be extended from Brazil to its neighbours to the south.

Similar solutions are also being sought to provide cross-border links to Brazil from Uruguay and Paraguay.

4 Large-Scale Testbed Networks in Brazil

Beginning in 2002, Brazil began to set up its own infrastructure for experimental research and development (R&D) in network technologies and distributed applications. So far there have been a number of separate initiatives.

4.1 Project GIGA Optical Testbed

Project GIGA is an ongoing project involving RNP and the CPqD Foundation in Campinas, which is the successor of the former state telecommunications monopoly's R&D laboratories. The original partnership was brokered by MCT, and led to the joint proposal of an optical testbed network, which was funded between 2003 and 2007 by the federal National Fund for the Technological Development of Telecommunications (FUNTTEL). Using fibres lent without cost by four separate telcos, a 750 km DWDM network was established by 2004 in southeastern Brazil (states of Rio de Janeiro and São Paulo), with nodes in 7 cities and access

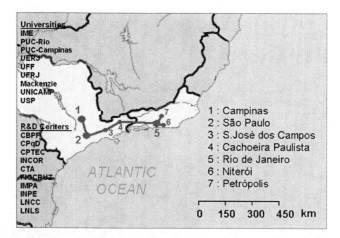

Fig. 5 Geographical location of the Project GIGA optical testbed (states of RJ and SP) (Courtesy: CPqD and RNP)

to laboratories in around 20 universities and research centres in this region (see Fig. 5). Initially 2.5G optics was used, but later this was partially upgraded to 10G optics. Ethernet technology has been used at level 2, both 1 and 10GigE.

All the optical transmission equipment used in this testbed has been provided by the Brazilian firm, Padtec, partially controlled by CPqD.

In addition to setting up the testbed, the funding covered the cost of extensive R&D activities, some coordinated by CPqD, in the areas of optical technologies and telecommunications applications, and the others by RNP in network protocols and distributed applications. In all, consortium-based R&D activities were carried out at more than 50 research institutions throughout Brazil, and validated on the testbed.

In addition to funded R&D activities, the testbed has also been used for providing high-capacity access to a small number of scientific groups with international collaborations, especially in grid computing and high-energy physics (HEP), in those cases when the previous connectivity proved insufficient. Thus, since 2004 the HEP group at the state university of Rio de Janeiro (UERJ), has participated in the bandwidth challenge (BWC) at the annual Supercomputing conferences in the US, using transmission facilities of the testbed between Rio de Janeiro and São Paulo.

R&D activities associated with Project GIGA activities have generated a very large number of products, including publications, of different kinds. See (Scarabucci 2005) for an early report on the project. The results of RNP-coordinated subprojects were presented at the (final) R&D Workshop for Project GIGA/RNP, held in September, 2007 (GIGA 2007).

A second round of FUNTTEL funding has been made available to CPqD for the so-called Phase 2 of Project GIGA, starting in 2009. RNP continues to collaborate with CPqD in these activities, which will be described later on in this report.

4.2 KyaTera TestbedNetwork

In 2003, the São Paulo Foundation for Research Support, FAPESP, launched the R&D programme, TIDIA (Information Technology for the Development of the Advanced Internet), of which one of the component projects was an optical testbed network, called KyaTera, which has been operating since 2007. This testbed links together research laboratories in institutions in 9 cities, using similar network technology to Project GIGA (Ethernet/WDM), and provides them with access to international connectivity in the city of São Paulo (see Fig.6). Further information about this network, can be found at (KyaTera 2010).

Again the research activities are mixed, with emphasis on optical technologies and distributed applications.

4.3 PlanetLab

The first PlanetLab (Peterson 2003) node in Brazil was created in Belo Horizonte in 2003, by a former student of Professor Larry Peterson of Princeton University.

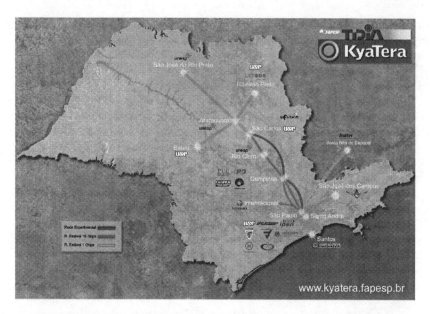

Fig. 6 The KyaTera testbed network in São Paulo state (Courtesy: FAPESP)

The following year, with a donation from the Intel Corporation, RNP deployed a further 3 nodes, and began to administer them on behalf of the networking research community in Brazil. Further interest in this facility has been generated over the years through occasional visits of PlanetLab specialists to the country, most recently from the European OneLab project in 2009.

4.4 GLIF – Global Lambda Integrated Facility

GLIF has worked steadily since 2003 to disseminate the adoption of hybrid packet-circuit networks, with the use of end-to-end circuits for high-volume flows – an early rationale for this was provided in (de Laat 2003). After gaining initial experience with such circuits in October 2007 for transmitting standard definition (SD) and compressed high definition (HD) video streams between Rio de Janeiro and Barcelona as part of the ArtFutura event, RNP, accompanied by its partners, ANSP and CPqD, joined the GLIF community in 2008, registering the SouthernLight GOLE, the Ipê network and the GIGA and KyaTera testbeds as usable GLIF resources. These appear on GLIF maps made available in that year (GLIF 2009).

GLIF community resources were used extensively in the inauguration of the first of the two 10Gbps links in July, 2009, with the worldwide première of the full length Brazilian feature film, *Quando a Noite Chega* (When Night Falls), made using 4K digital technology and simultaneously exhibited locally in São Paulo, and remotely at UCSD in San Diego (US) and at Keio University in Yokohama (Japan), to which the film was transmitted in compressed form (400 Mbps).

The event also featured uncompressed HD videoconferencing with Keio University (900 Mbps). In the aftermath of this event, RNP accepted the invitation to become a network member of the CineGrid community (CineGrid 2009).

4.5 Dynamic Circuit Provisioning

The GLIF community has been investigating dynamic provisioning of end-to-end circuits for several years, as a means of automating the hitherto manual process of setting up the circuits. Based on work carried out from 2005 in the DRAGON testbed at MAX (Lehman 2006) and the On-demand Secure Circuits and Advance Reservation System (OSCARS) system since 2004 at ESnet (Guok 2006), Internet2 and ESnet jointly deployed a pre-production dynamic circuit (DC) network in 2007 (Lehman 2007). It was thus demonstrated that what had previously required long setup times for circuits crossing multiple domains could now be performed in a matter of minutes, greatly reducing the operating costs and increasing the utility of this network service.

This demonstration has given new life to the GLIF activities, leading to the fusion of the previous Technical and Control Plane working groups, which had formerly been seen, respectively, as the present and future faces of GLIF. In addition, GLIF community influence is also visible in the effort that has recently been invested in standardisation of end-to-end circuits within the Open Grid Forum (OGF), through the working groups on Network Service Interface (NSI) and Network Markup Language (NML) (OGF-NSI 2008).

RNP's future backbone network in 2010 is also intended to adopt a hybrid packet-circuit architecture, and work is underway since 2008 on the design and implementation of a dynamic circuit capability, which will be able to interoperate internationally with other research networks. This project, known as FuturaRNP, is investigating a number of technical alternatives, such as DRAGON/OSCARS, UCLP/ARGIA (UCLP 2010) and AutoBAHN (GEANT2 2010), with the participation of research groups from 10 institutions, including CPqD.

The FuturaRNP project includes a testbed for development, which we call the Creeper Network (Rede Cipó), implemented as a Virtual Private LAN Service (VPLS) interconnecting the dedicated level 2 (Ethernet) networks at each institution (RedeCipó 2009). These single domain networks are similar to the "pods" used by Internet2/MAX for carrying out training in DC technology at Internet2 events (I2-DC-workshop 2008). By mid-2010 it is hoped to be able to migrate the resulting solution to the new Ipê network to be deployed around that time.

5 Experimental Future Internet R&D

RNP continues to partner CPqD in Phase 2 of Project GIGA, which is, by agreement, concentrating its activities on R&D in "Future Internet" architectures and applications. In addition to the original testbed in southeast Brazil described above, it is expected that the geographical coverage of the Phase 2 testbed will be extended to coincide with the greatly expanded 10Gbps core of the future Ipê network.

Large-scale Future Internet testbeds are beginning to be deployed in North America, the EU, Japan and Korea. In the USA, NSF launched its GENI (Global Environment for Network Innovations) programme in 2005 and, after several years spent on design, began in 2008 to formulate and deploy an experimental facility to support R&D into new network architectures (GENI 2009). The EU launched its FIRE (Future Internet Research and Experimentation) programme, also in 2008, based initially on a number of existing testbed projects: OneLab, Panlab, FEDERICA and Phosphorus (FIRE 2009). Meanwhile, in Japan the AKARI project was launched to design a New Internet by 2015 (AKARI 2008).

There are ostensibly several similarities between these different proposals, especially in the technologies adopted. In principle, all these testbeds seek to support simultaneous use by concurrent projects (architectures). To carry this out, extensive virtualisation is carried out, both of network resources, including switches, and of processing and storage devices available on the network. This latter facility was originally included as a fundamental part of PlanetLab technology, and this has now been extended into network virtualization by variants of PlanetLab, such as VINI, which enable the virtualization of a level 3 router based on a PC (Bavier 2006).

The most general model is that of GENI, which supposes the existence of a level 2 transport service linking network nodes containing programmable and virtualisable routers, as well as processing and storage elements. Among the programmable routers, apart from the VINI model, are such designs as OpenFlow (OF) and NetFPGA (McKeown 2008). On the other hand, the FEDERICA project has adopted the use of production IP routers which support router virtualization (FEDERICA 2009).

One thing is quite clear: there is considerable interest in interoperation of these different testbeds, leading to collaboration around the globe. In Brazil, several invitations have been received to participate in testbed projects which were proposed to GENI in 2009. Therefore, in the planning of a Brazilian Future Internet experimental facility, future interoperation with foreign partners is of great importance.

It should be mentioned that a couple of Brazilian Future Internet R&D projects are already underway: Horizon and WebScience.

Horizon is a project to study new Internet architectures, which is being jointly carried out by a consortium of French and Brazilian universities, together with industrial partners, and funded by their respective governments (Horizon 2010).

Web Science is a large consortium of more than 100 researchers from several leading universities, which is being funded for 3 to 5 years of research activity by CNPq, under its National Institutes of Science and Technology programme (WebScience 2010). RNP and a group of researchers from 5 universities have included in this project the establishment of a VINI-style testbed for experimental research into Future Internet architectures.

Lastly, interest has been expressed at government level in coordinating officially funded projects in the Future Internet area between Brazil and the EU, with a first call expected to be published in 2010.

References

1. AKARI project, New Generation Network Architecture: AKARI Conceptual Design, ver1.1 (2008), http://akari-project.nict.go.jp/eng/concept-design/AKARI_fulltext_e_preliminary.pdf (accessed on 15/12/2009)
2. Bavier, A., Feamster, N., Huang, M., Peterson, L., Rexford, J.: In VINI Veritas: Realistic and controlled network experimentation. In: Proc. ACM SIGCOMM, September 2006, pp. 3–14 (2006)
3. CineGrid (2009), http://www.cinegrid.org/ (accessed on 15/01/2010)
4. de Laat, C., Radius, E., Wallace, S.: The Rationale of Optical Networking. In: Future Generation Computer Systems, vol. 19, pp. 99–1008. Elsevier, Netherlands (2003), http://datatag.web.cern.ch/datatag/papers/fgcs2.pdf (accessed on 15/12/2009)
5. FEDERICA (2009), http://www.fp7-federica.eu/ (accessed on 15/12/2009)
6. FIRE, Future Internet Research and Experimentation (2009), http://cordis.europa.eu/fp7/ict/fire/ (accessed on 15/12/2009)
7. GEANT2 (2010), http://www.geant2.net/server/show/nav.756 (accessed on 16/12/2009)
8. GENI, Global Environment for Network Innovations (2009), http://www.geni.net/ (accessed on 15/12/2009)
9. GIGA, R&D Workshop for Project GIGA/RNP, LNCC, Petrópolis, RJ, Brazil. Conference agenda and record (September 2007), http://indico.rnp.br/conferenceDisplay.py?confId=33 (accessed on 13/12/2009)
10. http://www.glif.is/publications/maps/ (accessed on 15/12/2009)
11. Guok, C., Robertson, D., Thompson, M., Lee, J., Tierney, B., Johnston, W.: Intra and Interdomain Circuit Provisioning Using the OSCARS Reservation System. In: 3rd International Conference on Broadband Communications, Networks and Systems, BROADNETS 2006 (2006)
12. Horizon (2010), http://www.gta.ufrj.br/horizon/ (accessed on 15/01/2010)
13. I2-DC-workshop, Dynamic Circuit Network Hands-On Workshop 2008, http://dragon.maxgigapop.net/twiki/pub/DRAGON/Internet2DCNWorkshopJul2008/GMPLS_Hands-On_Workshop.ppt (July 2008)
14. KyaTera (2010), KyaTera website, http://www.kyatera.fapesp.br/ (accessed on 27/4/2010)
15. Lehman, T., Sobieski, J., Jabbari, B.: DRAGON: A Framework for Service Provisioning in Heterogeneous Grid Networks. IEEE Communications Magazine 44(3), 84–90 (2006)
16. Lehman, T., Yang, X., Guok, C., Rao, N., Lake, A., Vollbrecht, J., Ghani, N.: Control Plane Architecture and Design Considerations for Multi-Service Multi-Layer, Multi-Domain Hybrid Networks. In: INFOCOM 2007, IEEE (TCHSN/ONTC), May 2007, pp. 67–71 (2007)
17. McKeown, N., et al.: OpenFlow: Enabling Innovation in Campus Networks. ACM SIGCOMM Computer Communication Review 38(2), 69–74 (2008)

18. Open Grid Forum, Draft Charter for NSI-WG (August 2008), http://www.ogf.org/OGF24/materials/1390/NSI_charter_01.doc (accessed on 15/12/2009)
19. Peterson, L., Anderson, T., Culler, D., Roscoe, T.: A Blueprint for Introducing Disruptive Technology into the Internet. ACM SIGCOMM Computer Communication Review 33(1), 59–64 (2003)
20. Rede Cipó (a testbed for automatic provisioning of circuits) (2009), http://wiki.rnp.br/pages/viewpage.action?pageId=26969360 (accessed on 15/12/2009)
21. Scarabucci, R., Stanton, M., et al.: Project GIGA – High-speed Experimental Network. In: First International Conference on Testbeds and Research Infrastructures for the DEvelopment of NeTworks and COMmunities (TRIDENTCOM 2005), Trento, Itália, pp. 242–251 (February 2005)
22. Stanton, M.: Non-Commercial Networking in Brazil. In: INET 1993, San Francisco, Proceedings, San Francisco, Internet Society (August 1993), http://www.ic.uff.br/~michael/pubs/inet93.ps
23. Stanton, M., Ribeiro Filho, J., Simões da Silva, N.: Building Optical Networks for the Higher Education and Research Community in Brazil. In: 2nd IEEE/Create-Net International Workshop on Deployment Models and First/Last Mile Networking Technologies for Broadband Community Networks (COMNETS 2005), Boston, MA, USA, vol. 2, pp. 1499–1505 (2005)
24. Stöver, C., Stanton, M.: Integrating Latin American and European Research and Education Networks through the ALICE project. In: LANOMS 2003, Foz do Iguaçu, Proceedings, Foz do Iguaçu, UFPR (August 2003)
25. UCLP (2010), http://www.uclp.ca/ (accessed on 15/01/2010)
26. WebScience (2010), http://webscience.org.br/wiki (accessed on 15/1/2010)
27. WHREN-LILA 2010, http://www.whren-lila.net/ (accessed on 15/1/2010)

Description of Network Research Enablers on the Example of OpenFlow

Julius Werner

Technische Universität Berlin, Germany
jwerner@cs.tu-berlin.de

Abstract. This chapter describes OpenFlow, a specification developed by a research group at Stanford University that is proposed to be implemented by commercial switches and routers and would allow remote control of their forwarding behavior. It is aimed at providing researchers with an inexpensive and flexible platform to experiment with new network protocols on production-scale traffic. OpenFlow is further compared to two other projects that aim to enable network research, but differ totally in approach: the PlanetLab and the eXtensible Open Router Project (XORP). Finally, the NOX network operating system is described as an example for a project using OpenFlow's successful network hardware abstraction concept to implement a larger network management system.

1 Introduction

The Internet is an ever growing success and it does not look like that is going to slow down anytime soon. More than 500 million hosts (Comer 2008) and 30,000 autonomous systems (Bates et al. 2009) are connected today and more are sure to come. But at the same time, the Internet is growing old: more than 30 years have passed since advancements in early TCP development led to the specification and implementation of the one network layer protocol that would conquer the world (Postel 1978). IP started out as part of an experimental research project to provide interconnection between several computer networks of the time, which were more heterogenous and had far fewer hosts than today's systems. The precurring TCP implementations had gone through several major versions and quite a few important changes (Clark 1988). Similarly, some ideas for the new protocol were raised and some dropped again before the final draft – for example, the often criticized 32-bit address size might have been little more than coincidence (compare proposal for variable address length in Postel 1977). However, once the specification was implemented, it turned out to be good enough to warrant no serious further adjustments.

In the following years the Internet rapidly grew and eventually integrated commercial ISP networks. The Internet Protocol Suite was implemented many times on different systems, which dramatically increased the amount of coordination and work that would be necessary to implement further protocol changes. With the transition from the central NSFNET Backbone to a privatized, distributed backbone architecture in 1995, the Internet had finally become completely decentralized, further inhibiting architectural changes through the lack of a central coordinating instance that could propose and enforce them. The effects of this have long been evident: the most recent specification for the next generation network layer protocol has been lying in the drawer since 1998 (Deering and Hinden 1998) and while it has been slowly integrated into major host operating systems during the last decade, its adoption in the Internet backbone and regional ISPs is still far from global, with only a few isolated networks available. It may well be that IPv6 is once again already outdated before it is actually adopted, for some suggest it still includes too many of IPv4's mistakes.

In addition to the network layer itself, these difficulties also apply to exterior gateway routing: the Border Gateway Protocol has emerged as the de facto standard for interdomain routing and any changes to that which are not backwards-compatible would have to be conducted in the whole Internet at once. The situation here seems even worse: today's routing tables are bursting with entries, having to cope with hundreds of thousands of prefixes (Bates et al. 2009). While hardware vendors keep trying to increase their routers processing power to manage them, this is actually a design issue that should be solved on the whiteboard. But despite the widespread criticism of the protocol, truly innovative changes have low chances to be adopted by the Internet community (Carl and Kesidis 2008). Not even a standardized succession candidate (as with the Internet Protocol) has been decided upon yet. The National Research Council (2001) describes this process as ossification – the reluctance to replace widespread technology with something better, because the industry is not motivated to implement, let alone develop, changes with high cost and little immediate gain, which is aggravated by the fact that pioneering the change provides no benefit until most of the competitors have joined in. Unless a good incentive to conduct individual improvements can be found, the Internet might actually need to burst apart before the long overdue corrections can be implemented.

However, before anything can be changed, there has to be a good proposal for it. It has always been the research communities' part to design these innovations and while the IPv6 adoption may be slow, it at least shows that steady spreading of a new protocol can be possible as long as it is generally accepted and agreed upon. Reaching widespread acceptance is a long road, though, that involves not only innovative ideas and many refinement iterations, but also significant testing of all aspects of the new architecture, to convince the involved parties that it will hold its promises and be worth their money. Conducting those tests is difficult, however: they must be run at sufficient scale and with realistic traffic in terms of content as well as distribution and size. With network layer and routing innovations, this requires powerful routers that can cope with backbone level traffic, a requirement usually only fulfilled by hardware-accelerated commercial network devices. But implementing experimental new protocols in these is generally not possible, since they are closed systems with proprietary firmware that allows only the range of network operations required by usual administrators. Their operating systems cannot be extended or replaced because the vendors keep their secrets under close guard and do not disclose their architectures to the public. Although there have been efforts to create dedicated research hardware that allows total control of the packet processing, that tends to be too expensive for small research groups to acquire a sufficiently large network segment. This greatly inhibits testing and demonstration of experimental network and routing protocols.

Among the most promising efforts to mitigate this problem is the OpenFlow project. It aims at enabling large-scale network protocol and routing research on the regular commercial network hardware that is usually deployed at universities and research centers. This is done by merely defining a standard for controlling packet switching decisions – it is then the hardware vendor's part to implement it in their devices. The following will describe how OpenFlow-enabled switches make their forwarding decisions and how they can be configured to offer (almost) any desired functionality.

2 Background

The basic idea of OpenFlow (McKeown et al. 2008) is to create a system that grants researchers the greatest possible amount of control over the packet flow in their routers and switches, while still being easy and cheap to integrate into commercial networking hardware. In order to achieve this, packet handling decisions are based on a common subset of the information that different switches extract from a packet during its processing: the interface it was received on and the most basic contents of its Ethernet, IP and TCP/UDP header (when they are applicable). OpenFlow-enabled devices store tuples of this data in their 'Flow Table' and associate them with an action, e.g. dropping the packet or sending it out on a specific interface. For further flexibility, the Flow Table keys may contain wildcards, so that "Send all packets from any interface with VLAN ID 10 (taken from Ethernet header) and destination port 80 (taken from TCP/UDP header) out on interface 20." would translate to a valid Flow Table entry. An illustration of an example network including a switch with such an entry can be seen in Figure 1.

One of OpenFlow's goals is that researchers can conduct experiments right on the existing production hardware of their campus networks. This requires a strict separation of experimental and production traffic in the processing rules of a switch. To minimize the impact on existing production networks, the OpenFlow specification optionally allows the "Forward this packet through the switch's normal processing pipeline." action in Flow Table entries. This allows network administrators to run their OpenFlow-enabled switch with a default rule using this action for every packet and then define exceptions for special experiments, which might be recognized by a certain VLAN ID or EtherType value and handled differently. One should note that even if this action is not supported, the Flow Table concept itself could easily be used to simulate the default behavior of a switch or router.

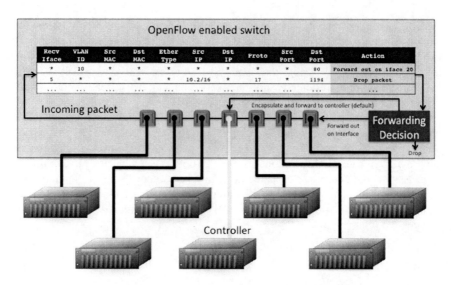

Fig. 1 OpenFlow enabled switch with Flow Table. One of the attached devices acts as controller.

In order to manage the Flow Table, OpenFlow defines a secure (SSL-based) network protocol that connects the OpenFlow-enabled switch or router with a controller. The controller is a server that can remotely add or remove entries from the switch's Flow Table. However, in order to dynamically manage flows the communication has to go both ways: the controller has to be informed about unhandled packets for which a routing decision has yet to be made. This is achieved by another Flow Table action: "Encapsulate this packet and send it to the controller." This facility makes it possible to test even complex new routing protocols on any OpenFlow-enabled switch, because all the control packets can be forwarded to and processed at the controller, which generates Flow Table entries from them and programs the switch accordingly. The net effect is a combination of the programmability and conceptual limitlessness of general purpose computers with the linerate pure processing power of commercial, enterprise-level networking hardware.

To complete the circle, the controller can encapsulate arbitrary packets and have the switch send them out on a chosen interface. One controller can operate multiple switches for a true centralization of the network logic (while still keeping the processing workload in the switches!), but a switch may also receive commands from multiple controllers for load balancing and redundancy. As the implementation and functionality of the controller is not constrained beyond compatibility with the protocol's instructions, controllers can have arbitrary complexity. Over time, powerful frameworks that multiplex the experiments of many researchers with different access rights onto the same network might be developed and even the concept of a comprehensive, distributed 'operating system of the network' becomes possible (see section 4).

The OpenFlow specification is developed and maintained by the OpenFlow Consortium – an open group of researchers and administrators of campus networks. The specification consists only of the OpenFlow network protocol (and the resulting Flow Table layout) itself, because the implementation on the network hardware will be closed and vendor specific. The Consortium therefore has to rely on vendors willing to adopt their concept and add OpenFlow-compatibility to their operating systems. However, due to the small and commonly used set of elements a Flow Table key consists of, the developers are confident that most of the existing target hardware could be OpenFlow-enabled with just a firmware upgrade and several vendors already added OpenFlow to their devices on an experimental basis (Naous et al. 2008). Right now, the OpenFlow specification is still in a late draft phase and does not yet advise widespread implementation in network hardware, because it may still be subject to change (Heller 2008).

3 Comparison to Other Approaches

The need for realistic network testing environments which produce significant results is not new and many approaches were and are currently being developed to fulfill that demand. These approaches often differ greatly in assumptions and target use cases, which results in many dissimilar ideas, although most of them are useful in a certain situation. It would be wise for a network researcher to consider each of the testing methods and environments, only three of which are outlined here, thoroughly for their suitability to his/her project.

Within this diversity, OpenFlow represents a compromise that tries to provide realistic test environments while staying cheap enough to be realizable. It creates a testbed in an existing single, confined network or even just part of a network while being able to simultaneously keep up the usual network service without impairment. Its most important advantage is that once OpenFlow-enabled firmware upgrades are readily available, network administrators can easily turn their existing campus networks into OpenFlow testbeds. This requires low costs from the hardware vendors and none from the users. But to achieve this ease of deployment, OpenFlow has to make some trade-offs: while packets can be switched at line rate, the forwarding decision is limited by the information available through the Flow Table. While packets can be arbitrarily processed by forwarding them to the controller, the resulting overhead is so large that this must be limited to a small number of packets (like a new routing protocol or a small stream of experimental data within a production network). And the conceptual limit to a single, local network makes OpenFlow unfeasible for realistic test deployments of global-scale protocols. It is certainly possible to send traffic between multiple independent OpenFlow networks over today's Internet, but in order to test routing protocols every hop on the path has to play its part – and the chances of convincing several Tier 1 ISPs to grant researchers access to their routers, even through the traffic separation of OpenFlow, seem slim.

3.1 PlanetLab

PlanetLab (Peterson et al. 2003) is a single, global-scale testbed which is maintained by the PlanetLab Consortium, a group of academic, industrial and government institutions that each have to run at least two nodes in PlanetLab as part of their membership. Each of these nodes offer an amount of resources like network bandwidth, processing power and memory, which can be used to instantiate virtual machines running a pre-installed but extensible Linux 2.4 kernel. What makes PlanetLab special is the concept of 'distributed virtualization': a user reserves resources on multiple nodes at once, called a slice, which are interconnected through a network overlay. This way, a whole distributed network can be reserved for an experiment, even though many other such experiments may run on the same nodes simultaneously through virtualization. Another interesting aspect of PlanetLab is that the management system itself is just another service running on a slice. With this approach, the PlanetLab core system provides only very basic management functions (like user authentication) and all higher level organization can be done transparently, interchangeably and with several independent parallel services – a concept the designers call 'unbundled management'.

Since its announcement in 2002, PlanetLab has grown significantly in size and accumulated more than 1000 nodes at nearly 500 sites (PlanetLab 2009). In addition many experimental services have been deployed over it, including routing overlays, content distribution networks, network embedded storage and QoS overlays to name only a few (Bavier et al. 2004). Many of these have been so successful that they basically developed into production services – thus PlanetLab has become not only a research but also a deployment platform, which was partially anticipated by its designers (Peterson and Pai 2007).

PlanetLab's greatest strength lies in its size: it is a viable platform to run global routing experiments or test services meant for planet-wide interaction. Because of the nodes global distribution, a PlanetLab node is always just a few hops away in most parts of the world – if the nodes are programmed to act as gateways, one can actually offer the whole world to take part in the experiment. The functionality is unlimited due to the researchers having full control of the processes running on each node in their slice. Its greatest weakness, however, is also its size: building and maintaining all the nodes is expensive. The Consortium's institutions need to pay continuously for the research possibilities offered, which is probably the reason they only grant access to their own researchers. Another weakness arises from the virtualization and overlay concepts: the significance of performance and QoS testing is limited, because the available processing power has to be shared with other slices and the packets are tunneled through the normal Internet with all its unpredictable behavior and traffic spikes.

Compared to OpenFlow, PlanetLab is targeted at different types of network research. OpenFlow is designed to research network or transport layer innovations in a realistic setting with commonly used hardware and production size traffic. PlanetLab's intent is on testing application or service layer innovations on a global scale, with more focus on functionality and scaling than performance. The two meet at the evaluation of new routing protocols, where OpenFlow is better suited for interior gateway protocols targeted at single networks and PlanetLab for global-scale exterior gateway protocols.

3.2 XORP

Another recent project intending to ease network research is the eXtensible Open Router Platform – in short: XORP (Handley et al. 2003). XORP is an open source software router implementation aimed at maximum support of existing routing protocols as well as total extensibility of every aspect of the router. This means that not only modules for new routing protocols, but also changes to the inner workings of the forwarding queue can easily be programmed and loaded into XORP. In addition, it is designed to be as robust as possible, including the separation of multiple loaded modules from the core package, so that bugs in one extension are less likely to crash the whole router. The designers included support for all functionality commonly expected today from an Internet core router. XORP can be compiled for most UNIX based operating systems and Microsoft Windows Server, but less supported systems might not be able to use all features. Although these operating systems usually bring their own routing software, it outperforms them by far in functionality as well as performance.

The XORP code itself focuses on providing an abstract, high-level API to its extensions and a rich library of state-of-the-art routing functionality and protocol support – the low level packet processing architecture is taken from the Click Modular Router (Kohler et al. 2000). Click is a modular packet forwarding framework that builds its forwarding information base in form of a graph connecting simple processing elements. Each element provides a single, simple processing function and the flow of a packet through several elements along the graph provides the compilation of those simple functions into a complex routing process. Click provides a remarkable processing performance: in a benchmark of the maximum loss-free forwarding rate on a 700 MHz Intel Pentium III, Click (v1.1) managed 333,000 packets per second, compared to a standard Linux kernel (v2.2.14) with 84,000 packets per second. This is mostly due to Click's DMA polling architecture, which can handle high loads faster than the hardware interrupt mechanisms usually employed by general purpose operating systems.

XORP is a great tool for early conceptual and functional testing of new routing protocols and algorithms. The PC platform and the extensibility approach offer limitless functionality and easy setup of development testbeds. However, a software-based router cannot possibly reach the performance necessary for line-rate processing of the production traffic in a large network. Although XORP performs much better than the usual network stacks included in operating systems, it utterly succumbs to the pure forwarding power of hardware-accelerated commercial routers with their TCAMs returning forwarding entries in nanoseconds. Due to this limitation, XORP based routers cannot be used for large-scale performance testing of routing or QoS protocols. The modular design might allow some of the forwarding path architecture to be moved to hardware with little effort, but packet forwarding hardware is generally only found in commercial routers and switches, which do not allow or support the installation of custom operating systems and software. Therefore, until the availability of affordable routing hardware with an open architecture, hardware-accelerated XORPs are but a theoretical exercise.

This unfeasibility of an open, extensible routing system with line-rate performance is exactly what motivated OpenFlow. With OpenFlow researchers can use a simple off-the-shelf switch to run their experiments with more traffic than a XORP system could ever hope to manage. The disadvantage is that the possible functionality of OpenFlow is limited by the columns in the Flow Table and the small bandwidth of the controller channel. It would therefore be impossible to perform production-scale testing of a router for a totally new network layer protocol in an OpenFlow network, because every packet has to be switched according to its network layer address. Usually this is the IPv4 address and thus a Flow Table key, but for other protocols the switch cannot keep different packet destinations apart and forwarding every packet to the controller to make that decision is impossible due to bandwidth and controller processing power limitations. In contrast, a XORP router (with appropriate extensions) would be able to route that traffic at about the same rate as normal IPv4.

4 Further Use: NOX

OpenFlow had been envisioned and designed to enable network research and testing and has grown to become a very flexible but still affordable network hardware abstraction and management framework. Its success has led developers with non-experimental uses in mind to discover its functionality. Having control of a switch's forwarding decisions from a remote controller in a standardized pattern and decoupled from the hardware implementation offers totally new possibilities in the area of centralized, abstract network management. One project that has been built upon OpenFlow is the NOX network operating system (Gude et al. 2008). The term 'network operating system' is used to describe that NOX offers for the network what usual operating systems offer for the computer: operating systems provide abstraction layers for underlying resources like hardware peripherals and memory, thus offering applications an execution environment while multiplexing several applications to those underlying resources and managing their interactions. In a similar manner, NOX intends to provide abstraction layers for the underlying network infrastructure (i.e. the switches and routers), offering 'network applications' an execution environment, while implementing the same multiplexing and management concepts between them. An application is then able to work with high-level abstractions of network elements, like users and hostnames, while NOX transparently maps them to the underlying identifiers (in this example, IP addresses).

NOX runs as a single central logical instance in a network, although it may be distributed to multiple physical instances for scalability. These instances are servers running the NOX software, which acts as an OpenFlow controller to nearby routers and switches. In order to achieve consistent behavior across the network, NOX keeps its state information in a centralized database called the 'network view'. There the network topology is stored, including the abstractions like users

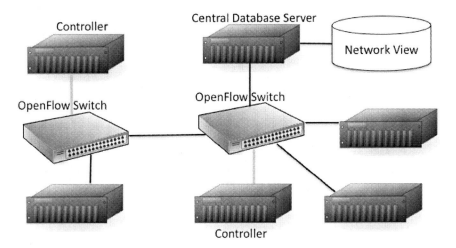

Fig. 2 Example setup. Note how every switch has its own controller, but state is stored centrally.

and hostnames as well as the underlying low-level information like MAC and IP-addresses. In addition, it might also be used to save global application state information, like policies to be enforced in the network. To create and update this network view, every NOX controlled OpenFlow switch forwards all packets from certain control protocols, like DNS, DHCP and RADIUS, to its controller, who in turn processes them to filter out information indicating a topology change, with which it updates the central database. An illustration of a possible (small) NOX network setup can be seen in Figure 2.

To support maximum scalability, NOX separates interaction with network traffic into a three layer model. The highest layer represents changes in the network view. These are determined by the controllers as described above and then, by means of the central database, propagated to the whole system. Topology changes are thus an expensive (in terms of overhead) operation, but stay manageable due to their rarity. On the middle layer are what the NOX developers termed 'flow initiations'. A flow generally represents a single, application specific end-to-end transmission, like a TCP connection, and can be identified by a Flow Table entry. The first packet of a flow, called flow initiation, is always forwarded to a controller. There its source and destination are determined and it may trigger NOX applications, which might decide that the flow should be denied, forwarded along its intended path or maybe rerouted transparently to a proxy server. When a path for it has been chosen, the controller adds appropriate entries in the affected switches' Flow Tables, so that any further packets for this specific flow can be processed by the switches themselves and don't need to bother the controller again. All necessary flow setup computations happen on a single controller, with no need to inform the system as a whole, which greatly contributes to scalability: should there be more flow initiations per second than the system can handle, one would simply add another controller server to the vicinity of the congested switch. The lowest layer of the NOX model represents all packets in the network. These are generally already part of a configured flow, so their processing happens solely inside the switches without interaction from a controller. This way, the bulk of the traffic is handled by fast, hardware-accelerated devices, with only the small fraction of them which constitute flow initiations ever being processed by the software-based controllers. This categorization of traffic is what enables NOX to utilize the processing performance of commercial network hardware, even though it is a distributed software system with centralized information and nearly unlimited, extensible functionality.

NOX itself only provides its architecture, abstraction layers and some useful basic library functions. Really utilizing its power requires running applications on top of it. Applications are simply C++ or Python programs that install event handlers in NOX. They run individually on every controller, using the NOX API to draw information from the network view (in addition to what is available from the event) and once again use the API to initiate Flow Table changes in all affected switches. A simple example application demonstrated in Gude et al. 2008 would hook into NOX' user authentication and whenever a new user authenticates to the system confine his flows to a specific VLAN. More advanced applications might communicate with each other and even between controllers to provide complex functionality. It is quite conceivable that with an appropriate system of applications, all of today's network management and configuration tasks could be moved to NOX and there automated, abstracted and condensed into nice, colorful GUIs. Imagine just plugging your hosts, switches and routers together and having NOX automatically configure the desired network architecture. Administrators could simply define the abstract policies for the network and lean back while NOX configures all devices to work together harmonically – and if a new switch needs to be added, it is plugged in and NOX extends the system's architecture and policies onto it automatically. The project will still have to go a long way to reach that spectrum of functionality, but its potential already looks very promising.

5 Conclusion

The OpenFlow specification is a very helpful innovation enabling researchers and network administrators to use the capabilities of their commercial network hardware to its fullest, not only in research applications but also in general network management. As a standardized interface to define packet processing rules remotely and independently from the hardware vendors, OpenFlow enables centralized network management at a scale that was formerly impossible, or only realizable through proprietary protocols and limitation on a single vendor. The critical question is whether hardware vendors would be willing to implement OpenFlow in the first place – but due to its current success, the 'OpenFlow-enabled' logo may quickly become a valuable selling proposition once the specification will actually be finished and released for general implementation.

Several other approaches to enable network research were and are currently being developed, but they complement rather than rival each other, either focusing on different phases of testing or on different fields of research. They might even end up being combined, as the OpenFlow developers expressed in their paper the hope that Stanford's OpenFlow network (and maybe others) would be integrated into the upcoming global-scale testbed GENI, which was inspired by PlanetLab.

Through the use of OpenFlow to shape the flows of a network, the already waning difference between routers and switches further decreases, since from OpenFlow's point of view they are the same, both offering a Flow Table to control their packet switching behavior. Proceeding further to whole networks controlled by a single, centralized network operating system, like NOX, the difference finally vanishes completely, because all former (interior) routing behavior is superseded by the Flow Tables.

References

1. Bates, T., Smith, P., Huston, G.: CIDR Report (2009), http://www.cidr-report.org/as2.0/
2. Bavier, A., et al.: Operating system support for planetary-scale network services. In: Proceedings of the 1st conference on Symposium on Networked Systems Design and Implementation, vol. 1, p. 19. USENIX Association, San Francisco (2004)
3. Carl, G., Kesidis, G.: Large-scale testing of the Internet's Border Gateway Protocol (BGP) via topological scale-down. ACM Trans. Model. Comput. Simul. 18(3), 1–30 (2008)
4. Clark, D.: The Design Philosophy of the DARPA Internet Protocols. In: SIGCOMM Symposium proceedings on Communications architectures and protocols, pp. 106–114 (1988)
5. Comer, D.: Computer Networks and Internets. Prentice-Hall, Englewood Cliffs (2008)
6. Deering, S., Hinden, R.: RFC 2460: Internet Protocol, Version 6 (IPv6) Specification (1998), http://www.ietf.org/rfc/rfc2460.txt
7. Gude, N., et al.: NOX: towards an operating system for networks. SIGCOMM Comput. Commun. Rev. 38(3), 105–110 (2008)
8. Handley, M., Hodson, O., Kohler, E.: XORP: an open platform for network research. SIGCOMM Comput. Commun. Rev. 33(1), 53–57 (2003)
9. Heller, B.: OpenFlow Switch Specification Version 0.8.9 (2008)
10. Kohler, E., et al.: The Click modular router. ACM Trans. Comput. Syst. 18(3), 263–297 (2000)
11. McKeown, N., et al.: OpenFlow: enabling innovation in campus networks (2008)
12. Naous, J., et al.: Implementing an OpenFlow switch on the NetFPGA platform. In: Proceedings of the 4th ACM/IEEE Symposium on Architectures for Networking and Communications Systems, pp. 1–9. ACM, San Jose (2008)
13. National Research Council, Looking over the fence at networks. National Academies Press, Washington D.C. (2001)
14. Peterson, L., et al.: A blueprint for introducing disruptive technology into the Internet. SIGCOMM Comput. Commun. Rev. 33(1), 59–64 (2003)
15. Peterson, L., Pai: Experience-driven experimental systems research. Commun. ACM 50(11), 38–44 (2007)
16. PlanetLab, An open platform for developing, deploying, and accessing planetary-scale services. PlanetLab Website (2009), http://www.planet-lab.org/
17. Postel, J.: IEN 66: TCP Meeting Notes, October 13 & 14 (1977), ftp://ftp.cs.tu-berlin.de/pub/doc/rfc/ien/scanned/ien66.pdf
18. Postel, J.: IEN 44: Latest Header Formats (1978), ftp://ftp.cs.tu-berlin.de/pub/doc/rfc/ien/scanned/ien44.pdf

Re-architected Cloud Data Center Networks and Their Impact on the Future Internet

Christian Esteve Rothenberg

University of Campinas (UNICAMP), Cidade Universitária "Zeferino Vaz"
Distrito de Barão Geraldo - Campinas - São Paulo, CEP 13083-852, Brazil
chesteve@dca.fee.unicamp.br

Abstract. Large-scale Internet data centers (DC) are empowering the new era of cloud computing, a still evolving paradigm that promises infinite capacity, no upfront commitment and pay-as-you-go service models. Ongoing research towards providing low-cost powerful utility computing facilities includes large-scale (geo)-distributed application programming, innovation in the infrastructure (e.g. energy management, packing), and re-thinking how to interconnect thousands of commodity PCs. In this chapter, we focus on the latter and review developments that are taken place in architecting data center networks (DCN) to meet the requirements of the cloud. Finally, we speculate on the potential impacts of such utility computing developments in shaping the future Internet by driving incentives of adoption of new protocols and architectural changes.

1 Introduction

In contrast to traditional enterprise DCs built from high-prize "scale-up" hardware devices and servers, cloud service DCs consist of low-cost commodity servers that, in large numbers and with appropriate software support (e.g. virtualization), match the performance and reliability of traditional approaches at a fraction of the cost. However, the networking fabric within the data center has not evolved (yet) to the same levels of commoditization [1]. Today's DCs use expensive enterprise-class networking equipment that require tedious network and IT management practices to provide efficient Internet-scale data center services. Consolidated on converged IP/Ethernet technologies, current DCNs are constrained by the traditional L2/L3 hierarchical organization which hampers the agility to dynamically assign services provided by virtual machines (VM) to any available physical server. Moreover, IP subnetting and VLAN fragmentation end up yielding poor server-to-server capacity even when relying on expensive equipment at the upper layers of the hierarchy [7].

Resource usage in the highly virtualized Cloud is very dynamic due to the nature of cloud services, causing unpredictable traffic patterns [9] for which common enterprise traffic engineering practices or intra-domain networks are not well suited and often result in over-subscription rates as high as 1:240 [6]. While not critical in enterprise networks, two main limitations of traditional Ethernet adversely affect its use in DCs: (1) scalability limits of ARP-broadcasting-based bridged spanning tree topologies; and (2) means to alleviate congestion without increasing latency. As a result, Ethernet-based store and forward switching potentially cause unacceptable high latencies in addition to dropped or reordered packets and excessive path failure recovery times even in the rapid versions of the spanning tree protocol (STP). An additional network management issue is concerned with the requirement of tweaking network path selection mechanisms to force the traffic across an ordered sequence of middleboxes (e.g. firewall, WAN optimization, Deep Packet Inspection, Load Balancer).

These and other shortcomings have made traditional Ethernet switching generally unsuitable for large-scale and high-performance computing needs of the cloud DCN. Industry efforts have been undertaken towards Data Center Ethernet extensions to provide QoS, enhanced bridging (IEEE 802.1 DCB), multipathing (IETF TRILL), Fibre Channel support, and additional Convergence Enhanced Ethernet (CEE) amendments. In the following, instead of delving into the market-driven incremental path of DC Ethernet solutions, we focus on the overarching requirements identified by industry and academia:

- Resource Pooling. The illusion of infinite computing resources available on demand requires means for elastic computing and agile networking. Hence, statistical multiplexing of physical servers and network paths needs to be pushed to levels higher than ever. Such degree of agility is possible (i) if IP addresses can be assigned to any VM within any physical server, and (ii) if all network paths are enabled and load-balanced.
- Scalability. Dynamically networking a large pool of location-independent IP addresses (i.e., in the order of millions of VMs) requires a large scale Ethernet forwarding layer. Unfortunately, ARP broadcasts, MAC table size constraints, and STP limitations place a practical limit on the size of the system.
- Performance. Available bandwidth should be high and uniform, independent from the endpoints' location. Therefore, congestion-free routing is required for any traffic matrix, in addition to fault-tolerance (i.e. graceful degradation) to link and server instabilities.

2 Re-architecting Approaches

Traditional DCN architectures consist of a tree of L2/L3 switches with progressively more specialized and expensive equipment moving up the network hierarchy. Unfortunately, this architectural approach is not only costly but results in the network becoming the bottleneck for cloud DC applications. Recent research in re-architecting DCNs has spurred creative designs to interconnect PCs at large, including shipping-container-tailored designs with servers acting as routers and switches as dummy crossbars [8] or re-thinking the flatness of MAC Ethernet addresses in favor of location-based pseudo MAC addresses [12].

The architectural approach of so-called next generation DCNs can be classified as server-centric or network-centric, depending on where the new features are implemented. The common goal is to provide a scalable, cost-efficient networking fabric to host Web, cloud and cluster applications. Many of these applications require bandwidth-intensive, one-to-one, one-to-several (e.g. distributed file systems), one-to-all (e.g. application data broadcasting), or all-to-all (e.g., MapReduce) communications among servers. Non-uniform bandwidth among DC nodes complicates application design (i.e. requires notion of data locality) and limits the overall system performance, turning the inter-node bisection bandwidth the main bottleneck in large-scale DCNs. The principal architectural challenges of DCNs are L2 scalability, limiting broadcast traffic, and allowing for multipath routing.

The rationale behind server-centric designs is to embrace the "end-host customization" and leverage servers with additional networking features. In a managed environment like the DC, servers are already commonly equipped with modified operating systems, hypervisors and/or software-based virtual switches to support the instantiation of networked VMs. Under a server-centric paradigm, routing intelligence is (sometimes solely) placed into servers handling also load-balance and fault-tolerance. Servers with multiple network interfaces act as routers (aka P2P networks) and switches do not connect to switches and act as crossbars. The approach is to leverage commodity hardware to "scale-out" instead of high-end devices to "scale up". The resulting server-centric interconnection networks follow the principles of e.g. mesh, torus, rings, hypercubes or de Bruijn graphs, well-known from the high performance computing (HPC) and peer-to-peer (P2P) fields.

Two remarkable examples from Microsoft Research branches are VL2 [6] and Bcube [8]. VL2 describes a large Virtual Layer 2 Ethernet DCN that builds upon existing networking technologies and yields uniform high capacity and traffic fairness by virtue of valiant load balancing (VLB) to randomize traffic flows throughout a 3-tiered switching fabric using IP-in-IP encapsulation and Equal Cost Multi-Path (ECMP). In order to support agility, VL2 uses flat addresses in the IP layer and implements address resolution (mapping of application IP address to location IP address) by modifying the end systems and querying a scalable directory service. Bcube [8] is a shipping-container-tailored DCN design where switches only interconnect servers acting as routers. Scalable, high-performance forwarding is based on source routing upon a customized shim header (additional packet header) inserted and interpreted by end-hosts, which are equipped with multiple-cores and programmable network interface cards (e.g. NetFPGA). Container-based modular DCs emerge as an efficient way to deliver computing and storage services by packing a few thousand servers in a single container. The notable benefits are the easy deployment (just plug-in power, network, and chilled water), the high mobility, the increased cooling efficiency, and foremost the savings in manufacturing and hardware administration. Challenges include high resilience to network and server failures, since manual hardware replacement may be unfeasible or not cost-effective.

On the other hand, network-centric designs aim at unmodified endpoints connected to a switching fabric such as a Clos network, a Butterfly or a fat-tree topology. For instance, the fat-tree topology is very appealing because it provides an enormous amount of bisection bandwidth (without over-subscription) while using only small, uniform switching elements [1, 4]. The key modification happens at the control plane of the network, leaving end hosts and the switch hardware untouched, exploiting the availability of an open API such as OpenFlow [11]. Network customization through switch programmability requires network-wide controllers to install the forwarding tables of switches, resolve IP identifiers to network locators in response to ARP requests intercepted at edge switches, which are programmed for the desired line-speed packet flow handling actions (e.g., header re-writings). For instance, PortLand [12] is a native layer 2 network based on translating Ethernet MAC addresses into position-based "pseudo" MAC addresses. Network equipment vendors have already begun building switches from merchant silicon using multi-stage fat-tree topologies internally [4].

If we abstract the details of proposed DCN architectures (see examples in Table 1), in addition to design for failure (breakdown of servers and switches assumed to be common at scale), the following design principles can be identified:

- Scale-out topologies. Similar to how HPC clusters have been using two and multi-layer Clos configurations for around a decade because of their nice properties (e.g., blocking probability, identical switching elements), scale-out topologies of cloud DCN commonly follow a 3-tier arrangement with a lower layer of top-of-rack (ToR) switches, a layer of aggregation switches, and an upper layer of core switches. However, as long as they offer large path diversity and low diameter, other scale-out topologies can be considered (e.g. DHT-like rings, Torus).
- Separating Names from Locations. Identifier-locator split is not only an issue of Internet routing research (cf. IRTF RRG, LISP [110]) to overcome the semantic overload of IP addresses, but is the common approach in DCNs to enable scalability and resource pooling of IP addressable services. The lack of topological constraints when assigning IP addresses to physical servers and VMs, enables cloud services to expand or contract their footprint as required. In this context, IP addresses are not meaningful for packet routing, which is commonly based on a revisited (usually source-routing-based) packet forwarding approach.
- Traffic randomization. The burstiness and the unpredictability of DC traffic patterns [9] requires routing solutions that provide load balancing for all possible traffic patterns, i.e., demand-oblivious load balanced routing schemes. Oblivious routing has shown excellent performance guarantees for changing and uncertain traffic demands in the Internet backbones and more recently in DCN environments [6, 8]. For instance, VLB bounces off every flow to random intermediate switches and can be implemented via encapsulation (e.g., IEEE 802.1ah, IP-in-IP) or revisited packet header bit spaces (e.g., position-based hierarchical MAC addresses [12], Bloom-filter-based Ethernet fields [16]).
- Centralized controllers. In order to customize the DCN and achieve the meet control requirements, a direct networking approach based on logically centralized controllers is a common approach to transparently provide the

networking functions (address resolution, route computation) and support services (topology discovery, monitoring, optimization). Implemented as fault-tolerant distributed services in commodity servers, centralized directory and control plane services have shown to scale well and be able to take over the network control, rendering flow-oriented networking, load balancing, health services, multicast management, and so on.

Table 1 Comparison of published architectural approaches for cloud data center networks.

	VL2 [6]	Monsoon [3]	Bcube [8]	Portland [6]	SiBF [16]
Topology	3-tier 5-stage Clos	3-tier 5-stage Clos	Hypercube	3-level fat-tree	Any
Routing & Forwarding	IP-in-IP encapsulation	MAC-in-MAC tunneling	Shim-header-based source routing	Position-based hierarchical MAC	Bloom-filter-based source routing MAC
Load balancing	VLB	VLB	Oblivious	Not defined	VLB
End-host modification	Yes	Yes	Yes	No	No
Programmable switches	No	Yes	No	Yes	Yes

3 Trends towards the Inter-cloud

Cloud DCs are like factories, i.e., the number one goal is to maximize useful work per dollar spent. Hence, many efforts are devoted to minimize the costs of running the large scale infrastructures [5], which requires bringing down the power usage effectiveness (PUE) levels and potentially benefiting from tax incentives for (near) zero-carbon-emission DCs. In this context, energy efficiency of photonic cross-connects outperform the electrical counterparts. However, before we assist to the first all-optical DCN, the price-per-Gbit of optical ports needs to sink at a higher rate than the electrical versions. Further technology market break even points that need to be monitored include high speed memory and solid state disks. Spinning-based hard disks offer the best bit-per-dollar ratio but are limited by their access time, which motivates the design of novel DC architectures [14] where information is kept entirely in low latency RAM or solid state flash drives, while legacy disks are deprecated to back-up jobs. Another ratio that may motivate the design of new (content-centric) inter-networking solutions is the memory vs. transit price, which may motivate DCNs (and routers) to cache every piece of data in order to reduce the costs of remote requests.

The so-called *green networking* trend favors connections to remote locations close to (cheap/clean) energy sources. Recent studies [15] in cost-aware Internet routing have reported 40% savings of a cloud computing installation's power usage by dynamically re-routing service requests to wherever electricity prices are lowest on a particular day, or perhaps even where the data center is cooler. Such green inter-networking approaches require routing algorithms that track electricity prices and take advantage of daily and hourly fluctuations, weighting up the physical distance needed to route information against the potential savings from reduced energy costs.

With the advent of *software-defined networks* as proposed by the OpenFlow initiative [13], data center networking can be tackled as a distributed software problem, tractable by the experienced community of distributed system developers contributing to the emergence of warehouse-scale computers. The OpenFlow protocol enables software-defined networks by specifying a standard way for controlling packet switching decisions in software while keeping the hardware vendors in charge of the device implementation. This separation of concerns leads to a promising combination of the programmability and conceptual limitlessness of general purpose computers with the line speed processing power of commercial networking hardware.

The following domains can be identified as distinctive areas of opportunities for *optical* technologies:

1) Intra-DCN with all-optical technology, potentially with multiple lambdas per port and WDM-based solutions. Innovation is call for to provide fast reconfigurable optical paths to circumvent congestions by dynamically setting up light paths between ToRs (cf. [18]), or novel configuration-less multicast-friendly optical switching, e.g., borrowing from the Bloom filter principle of the electrical domain (cf. [16]) to provide pure optical switching based on the presence of a certain combination of optical signal wavelengths.

2) Inter-DCN solutions to support the (live) migration of VMs (i.e., workload mobility) and data-intense computation jobs from the enterprise to the cloud and vice-versa, the so-called cloud-bursting. In addition to being bandwidth-hungry, cloud-bursting requires scalable networking solutions with built-in security and control mechanisms (aka Virtual Private Lan Services - VPLS) that provide addressing protocol and topology transparency over QoS capable virtual private clouds. In this context, multi-domain optical technologies may be an aid to the emergence of an Inter-Cloud, i.e., the inter-networking of Clouds (public, private, internal) for the dynamic creation of federated computing environments that promise to leverage the Internet to an even more consolidated global service platform.

Before the vision of the Inter-Cloud becomes real there are a number of challenges that need to be solved in a joint effort by all players in the Cloud. In a recent talk [3], Internet evangelist Vint Cerf has said that:

> *"The Cloud represents a new layer in the Internet architecture and, like the many layers that have been invented before; it is an open opportunity to add functionality to an increasingly global network"*

It can be argued that the emergence of the Cloud is another example of the history repeating itself. The obvious analogue is the advent of the fifth computing utility resembling the time when simple remote hosts logged into the main computers hosting the applications and the users data. Moreover, Cloud Initiatives that have an analogue in the Internet's past include (i) the rising importance of academia, (ii) the increasing interest in interoperability among cloud vendors, and (iii) the carrier interest in new service opportunities.

In a way, today's clouds like network islands before IP. The current cloud scenario can be compared to the lack of communication and familiarity that existed among computer networks in 1973. The lack of inter-cloud standards poses issues if users want to move their data from cloud A to cloud B. Interoperability is not only beneficial to users in order to avoid cloud provider lock-in and reduce the risk of service availability or data losses. Cloud providers may find the incentives for having multiple clouds interact with each other to take advantage of the joint resource pool offered through such combinations.

According to Vint Cert, some of the open questions on Inter-Cloud include:

- How should one reference another cloud system?
- What functions can one ask another cloud system to perform?
- How can one move data from one cloud to another?
- Can one request that two or more cloud systems carry out a series of transactions?
- If a laptop is interacting with multiple clouds, does the laptop become a sort of "cloudlet"?
- Could the laptop become an unintended channel of information exchange between two clouds?
- If we implement an inter-cloud system of computing, what abuses may arise?
- How will information be protected within a cloud and when transferred between clouds?
- How will we refer to the identity of authorized users of cloud systems?
- What strong authentication methods will be adequate to implement data access controls?

Web pioneer Sir Tim Berners-Lee has been working on powerful ideas that may solve the Inter-Cloud problems. As noted by Vint Cerf, the idea of data linking is well-suited to provide a part of the vocabulary needed to interconnect computing clouds. More precisely, the semantics of data and of the operations on the data, and the vocabulary in which these actions are expressed may be the beginning of an inter-cloud computing language.

The Inter-Cloud issues above belong to the new emerging application-level layer on top of the core Internet infrastructure. There are however other internetworking implications (beyond 40GE) driven by the emergence of the Inter-Cloud. Justified by the business aspects of the Cloud, the following protocols and network solutions can be identified as gaining incentives for adoption due to the inter-networking requirements of the plane-scale Cloud:

- IPv6 to provide unique addressing for millions of virtual machines
- Inter-domain IP Multicast
- IP Mobility
- Locator/ID Separation Protocol (LISP)
- DNS Security Extensions
- Secure BGP

• Novel Inter-domain routing control platforms (cf. Transit Portal [17])
A comprehensive review of protocols and formats for Cloud computing interoperability can be found in [16]. The architecture for Inter-Cloud standards is still very much in its early stages and the market implications are unclear. We may however expect that such a global market of cloud resources (e.g. computational, storage) will eventually emerge. The implications for the connectivity market are also unclear, but we may speculate that the geo-distributed data center footprint may drive the development of novel connectivity options and a cloud-oriented connectivity markets beyond traditional multi-homing.

4 Conclusions

We have reviewed the major architectural changes present in the new generation data centers empowering the new era of cloud computing. The research upfront is multi-disciplinary and has spurred creative designs beyond the traditional L2-L3 hierarchical IP/Ethernet network arrangements. The networking requirements of the Cloud are far more reaching than the data center internal designs. The so-called Inter-Cloud requires the development and adoption of a new set of inter-cloud protocols to ensure interoperability. Similarly, innovation in Inter-Cloud connectivity options is called for to blur the barriers between the geographical locations of the virtual machine instances in order to enable unfettered workload mobility. Finally, we have speculated on the potential impacts of such utility computing developments in shaping the future Internet by driving incentives of adoption of new protocols and architectural evolution.

Acknowledgments. This work is partly founded by Ericsson Research and would not have been possible without the support and discussions of Prof. Maurício Magalhães, Fábio Verdi and Rafael Pasquini. A short version of this chapter can be found in IEEE ComSoc Optical Networking Technical Committee (ONTC) Newsletter PRISM, Vol. 1. No 2. March 2010.

References

1. Al-Fares, M., Loukissas, A., Vahdat, A.: A scalable, commodity data center network architecture. SIGCOMM CCR 38(4), 63–74 (2008)
2. Bernstein, D., Ludvigson, E., Sankar, K., Diamond, S., Morrow, M.: Blueprint for the Intercloud - Protocols and Formats for Cloud Computing Interoperability. In: ICIW, Fourth International Conference on Internet and Web Applications and Services (2009)
3. Cloud Computing and the Internet,
   http://googleresearch.blogspot.com/2009/04/cloud-computing-and-internet.html (last access on 03/08/09)
4. Farrington, N., Rubow, E., Vahdat, A.: Data Center Switch Architecture in the Age of Merchant Silicon. In: IEEE Hot Interconnects, New York (August 2009)
5. Greenberg, A., Hamilton, J., Maltz, D., Patel, P.: The cost of a cloud: research problems in data center networks. SIGCOMM CCR 39(1) (2009)

6. Greenberg, A., Hamilton, J., Jain, N., Kandula, S., Kim, C., Lahiri, P., Maltz, D., Patel, P., Sengupta, S.: Vl2: a scalable and flexible data center network. SIGCOMM CCR 39(4), 51–62 (2009)
7. Greenberg, A., Lahiri, P., Maltz, D., Patel, P., Sengupta, S.: Towards a next generation data center architecture: scalability and commoditization. In: PRESTO 2008, pp. 57–62. ACM, New York (2008)
8. Guo, C., Lu, G., Li, D., Wu, H., Zhang, X., Shi, Y., Tian, C., Zhang, Y., Lu, S.: Bcube: a high performance, server-centric network architecture for modular data centers. In: SIGCOMM 2009. ACM, New York (2009)
9. Kandula, A., Sengupta, S., Patel, P.: The nature of data center traffic: Measurements and analysis. In: ACM SIGCOMM IMC (November 2009)
10. Locator/ID Separation Protocol (LISP), http://tools.ietf.org/html/draft-farinacci-lisp-10 (accessed March 2010)
11. McKeown, N., Anderson, T., Balakrishnan, H., Parulkar, G., Peterson, L., Rexford, J., Shenker, S., Turner, J.: Openflow: enabling innovation in campus networks. SIGCOMM CCR 38(2), 69–74 (2008)
12. Niranjan, N., Mysore, A., Pamboris, N., Farrington, N., Huang, P., Miri, S., Radhakrishnan, V.: Subramanya and Vahdat A Portland: a scalable fault-tolerant layer 2 data center network fabric. In: SIGCOMM 2009 (2009)
13. The OpenFlow Switch Consortium, http://www.openflowswitch.org/
14. Ousterhout, J., et al.: The case for ramclouds: Scalable high performance storage entirely in dram. SIGOPS Oper. Syst. Rev. 43(4), 92–105 (2010)
15. Qureshi, A., Weber, R., Balakrishnan, H., Guttag, J., Maggs, B.: Cutting the electric bill for internet-scale systems. In: SIGCOMM 2009 (2009)
16. Rothenberg, C., Macapuna, C., Verdi, F., Magalhães, M., Zahemszky, A.: Data center networking with in-packet Bloom filters. In: 28th Brazilian Symposium on Computer Networks (SBRC), Gramado, Brazil (May 2010)
17. Valancius, V., Feamster, N., Rexford, J., Nakao, A.: Wide-area route control for distributed services (revision pending). To appear in Proc. USENIX Annual Technical Conference (June 2010)
18. Wang, G., Andersen, D., Kaminsky, M., Kozuch, M., Ng, T., Papagiannaki, K., Glick, M., Mummert, L.: Your data center is a router: The case for reconfigurable optical circuit switched paths. In: Proc. of HotNets-VIII (2009)

Improving the Scalability of Internet Routing

Christian Vogt

Ericsson Research, California, US
christian.vogt@ericsson.com

Abstract. This chapter explores a scalability problem in Internet routing as well as potential solutions. The reader will understand why routing scalability on the Internet is worth improving, and what benefits and limitations existing solution proposals have.

1 Introduction

Networks at the edge of the Internet increasingly often switch from classic, provider-allocated IP addresses to provider-independent IP addresses. They do this to avoid internal renumbering when changing providers and, when multi-homed with multiple providers, to facilitate load balancing and fail-over between these providers. Unfortunately, the adoption of provider-independent IP addresses causes undesirably fast growth in the size and update frequency of the global routing table: Provider independence reduces the topological significance of IP addresses, and hence defeats the aggregatability of routing table entries. Load balancing and failover for such routing table entries consequently cause route changes Internet-wide. The fast growth in the size and update frequency of the global routing table strain memory and processing capacities in Internet core routers and require shorter and shorter time-to-upgrade intervals.

Mitigating the scalability problem of the Internet routing architecture is important to enable a continued efficient functioning of the Internet and reasonable upgrade intervals for Internet core routers. A scalability problem always becomes more and more significant as the affected system grows. The Internet routing scalability problem may have been acceptable in the early days of the Internet, when the Internet was still small. It may even still be acceptable today. But the problem is becoming more and more noticeable, and it will continue to. The need for a more scalable Internet routing architecture has consequently sparked a considerable body of research efforts throughout the recent past. Proposed solutions are diverse, ranging from backwards-compatible, evolutionary techniques, to revolutionary clean-slate approaches. Some solutions tackle the problem with more scalable router architectures, others advocate changes to addressing schemes and routing protocols.

This chapter explores the scalability problem in Internet routing and potential solutions. The reader will understand why routing scalability on the Internet is worth improving, and what benefits and limitations existing solution proposals have.

2 Evidence of Scalability Problem

The Internet routing scalability problem can best be evidenced with the growth of the global routing table that Internet core routers are required to maintain. Figure 1 depicts this growth. It was generated based on the data and tools available at Geoff Huston's CIDR Report web site [1]. Figure 1 shows the size of the global routing table, in 1000's of entries, as a function of time from 1989 until present. The size of the global routing table has more than doubled over last five years. This is more than twice the increase of the natural growth of the Internet itself.

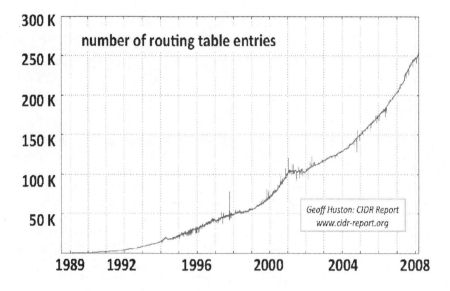

Fig. 1 Evidence of scalability problem

Routers will continue to support the increase in demand, but they will have to be upgraded more frequently. Accounts by network operators and router vendors suggest that the time-to-upgrade may reduce from once every seven years to once every four year, thus increasing network operators' capital expenditures.

With the size of the routing table grows the frequency of routing table updates. This, too, strains the capacities of Internet core routers and reduces the time-to-upgrade. Figure 2, generated based on the same source as figure 1, captures the number of updates to the global routing table per day, throughout the time from January 2005 to September 2006.

Accordingly, there is a steady growth in the frequency of routing table updates, an increase by two thirds over the period of observation. Aggravating the problem are considerable fluctuations in the update frequency, peaking at times at about a million updates per day. It has been shown that the origins of these update bursts are oftentimes only a few networks.

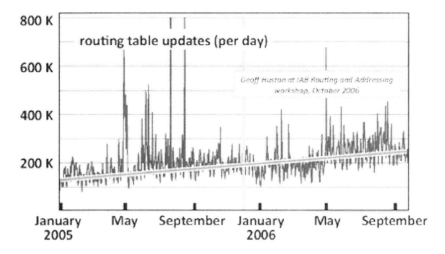

Fig. 2 Update frequency of global routing table

The increase in routing table entries and routing table upgrades are both driven by a local benefit, but have global cost. There are two reasons for this:

- Routing table entries and updates affect routers Internet-wide, and hence consume resources globally.
- Routing table updates in particular cause temporary routing instabilities until the routing system has converged on new routes that reflect an update. These instabilities can have global effect.

Like the energy resources of earth, the routing resources of the Internet are hence a common good that is exploitable by individual network operators for their own benefit. Problems of this kind are often classified as "a tragedy of the commons".

3 Routing and Addressing Recap

Traditionally, the Internet routing system differentiates between two types of networks: networks that carry transit traffic for other networks – so-called *providers* –, and networks that do not – so-called *edge networks*. Providers each get their own individual addressing space, like providers 1, 2, and 3 in figure 3 get the spaces denoted by address prefixes 1000::/32, 2000::/32, and 3000::/32, respectively. Then, providers use the Border Gateway Protocol, BGP, to tell each other which addressing spaces they use and for which other addressing spaces they provide transit service.

Each provider gathers the information received through BGP in a global routing table. For example, the global routing table of provider 1 in figure 3 would tell

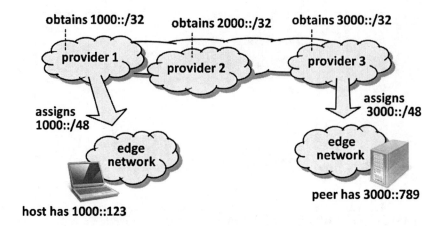

Fig. 3 Address space assignment from providers to edge networks

that addressing space 2000::/32 is neighboring to the right, and that packets destined to addressing space 3000::/32 must be sent the same way. The global routing table of provider 2 would point to the right for addressing space 1000::/32, and to the left for addressing space 3000::/48. Provider 3 would forward packets for addressing spaces 1000::/32 and 2000::/32 both via provider 2.

The preferred approach for edge networks is different: Edge networks are called upon to use addressing space allocated by their respective providers. The left edge network in figure 3 would get a slice of provider 1's addressing space, and the right edge network would get a slice of provider 3's addressing space. Accordingly, a host in the left edge network would configure an IP address that belongs to provider 1, and a host in the right edge network would configure an IP address from provider 3.

The reason for allocating provider addressing space for edge networks, and not giving edge networks provider-independent addressing space, is scalability: Edge networks become implicitly reachable via their respective provider's entry in the global routing table. They do not need their own routing table entry. In specialist terms, IP addresses used inside an edge network are called aggregatable with the addressing space of the edge network's provider. Thus, in the example of figure 3, packets destined to the host shown on the lower left of the figure find the destination edge network by following the routing table entries for provider 1, and provider 1 forwards the packets via the right border link.

Unfortunately, the use of provider-assigned address space in edge networks limits the flexibility of edge networks in two ways:

- Use of provider-assigned address space implies that, when an edge network becomes multi-homed, it will obtain addressing space from each of its providers, and hosts will configure IP addresses from each such addressing space. Packets that are sent from or destined to either of these IP addresses will then be routed via the provider to which the IP address belongs. Re-routing for the purpose of failover or load balancing is impossible.

Improving the Scalability of Internet Routing

- Use of provider-assigned address space causes a slight form of provider lock-in. Edge networks get new addressing space whenever they change providers, and this requires them to undergo renumbering. Renumbering is an expensive and time-consuming procedure that requires substantial manual efforts: It implies address changes to hosts and network equipment edge-network-wide. It affects routers, host, DNS and DHCP servers, firewalls, intrusion detection systems, remote-monitoring systems, load balancers, as well as scripts and configuration files.

It is consequently no surprise that edge network operators are reluctant to use provider-assigned address space. They prefer to have their individual address space, like providers have it.

Current practice on enabling multi-homing and eliminating renumbering is indeed for edge networks to obtain provider-independent addressing space – like the left and right edge networks in figure 4 are obtaining the address spaces denoted by address prefixes abc::/32 and def::/32, respectively. Accordingly, the IP addresses of hosts in these edge networks are no longer bound to provider 1 or to provider 2. Packets sent from such an IP address can be routed to either provider.

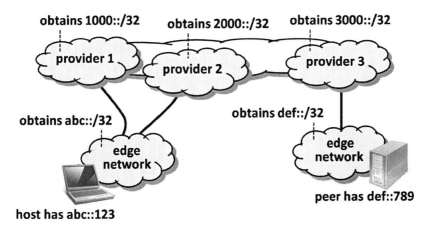

Fig. 4 Provider-independent addressing

On the other hand, provider-independent addressing is responsible for the Internet routing scalability problem. In order for packets to find their way towards a provider-independent IP address, the addressing space of the edge network needs to be advertised among providers via BGP. This implies higher router load, and thus less efficient packet forwarding:

- Increased routing table size: Global routing tables must list provider-independent edge network addressing space separately because that addressing space can no longer be aggregated with the provider's addressing space. The result is an increase in the size of global routing tables. This effect can be

substantial because the number of potentially multi-homed edge networks is orders of magnitude larger than the number of providers.
- More frequent routing table updates: For an edge network to change the provider via which packets will reach it, addressing space announcements in BGP must be updated. The result is an increase in the update frequency of the global routing table.

It is hence the selfish, yet legitimate interest of individual network operators that is fueling the Internet routing scalability problem, with an effect that spans the Internet globally.

4 Solution Approaches

The Internet routing scalability problem is not new; it has been a concern for almost the entire life of the Internet. But still, the Internet routing architecture has so far gone without substantial upgrades. Addressing has always been according to the Internet Protocol, and version 6 of the protocol does not differ conceptually from version 4. Routing has not changed since the introduction of the Border Gateway Protocol in the 1980's. There was only one evolutionary step, Classless Inter-Domain Routing, CIDR, which is a means to improve the efficiency of address allocation and address aggregation.

Since a few years, however, people are more seriously considering improvements to the Internet routing architecture – not only in the engineering community, but also and currently foremost in the research community. In October 2006, the Internet Architecture Board had a meeting on the routing scalability problem. The problem was brought up foremost by router vendors, who had decided to request the engineering community to do something about the problem. The Internet Architecture Board decided to investigate further into the severity of the problem and into possible solutions. It chartered a Routing research group inside the Internet Research Task Force for these efforts.

Since its instantiation, the Routing research group has been the main forum of all Internet routing scalability research. Most of the proposed solutions are of three basic classes:

- New router designs that can better handle the increased routing table size and update frequency. These aim at prolonging the time-to-upgrade for Internet core routers. An old example of a change in router design is compression of the forwarding table. This reduces lookup latencies and therefore helps routers cope with the growth of the routing table. It is not a complete solution to the routing scalability problem, though, because it fails to tackle the growth in the number of route updates. Therefore it cannot improve the stability of the Internet routing system either.
- Better address aggregation methods: Today, address aggregation in routing tables is based on the assumption that all or most of the addresses are provider-assigned. This assumption breaks with the adoption of provider-independent addresses, which in turn defeats address aggregation. Are there new methods

for aggregation that allow for provider independence? An example of a new address aggregation method is geographic aggregation. This reduces both the size and the update frequency of the global routing table. Geographic aggregation is controversial, however. While acceptable for some network operators, other network operators fear that the technique limits their abilities to express policy in the BGP.
- The class of address mapping solutions, which separate routing and addressing in the Internet core from routing and addressing in edge networks, and introduce a mapping between the two. The mapping can take place either in hosts or in the network, depending on the particular solution. Examples of host-based address mapping solutions are Shim6 and Six/One; examples of the network-based address mapping solutions are LISP and Six/One Router.

Address mapping has been a focus of the discussion in the engineering and research community, in particular in the Routing research group of the Internet Research Task Force. The following sub-sections will therefore look at this class of solution in more detail.

4.1 Shim6: Host-Controlled Multi-Homing

The probably the best-known host-based address mapping solution is Shim6 [2]. It has been developed in the Internet Engineering Task Force to enable multi-homing in a scalable way in IP version 6. The design rationale for Shim6, and for host-based address mapping solutions in general, is that the host is architecturally the best place to put multi-homing functionality: Only hosts have a full view of end-to-end connectivity and therefore can best decide which path to use.

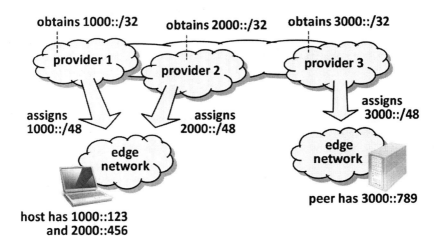

Fig. 5 Shim6

In Shim6, address mapping is inside the host, between the multiple addresses that a host is using: Shim6 uses provider-assigned addressing space inside a multi-homed edge network. Multi-homed edge networks therefore use multiple address spaces simultaneously. And hosts configure an IP address from the addressing space of each provider. A host can start a new communication session via either provider, and later on move this session to a different provider in case the path via the first provider fails.

Figure 5 illustrates the functioning of Shim6. The left edge network is assigned address spaces 1000::/48 and 2000::/48 by providers 1 and 2, respectively, and the host attaching to the left edge network creates an IP addresses 1000::123 and 2000::456 from these address spaces. Packets exchanged via IP address 1000::123 are then routed via provider 1, and packets exchanged via IP address 2000::456 are routed via provider 2. The host may start a communication session using either IP address. Upon a failure of the path via the original provider, the host can switch over to the other IP address, and thereby re-route the session via the alternative provider. To make the address change transparent to applications, the host and its peer perform an address mapping at their IP layers, such that the alternative IP address is only visible in packets only while on the network, and applications continue to use the original IP address.

A reason for criticism of Shim6 is that is does not enable the network operators to control though which provider a host's packets are routed. This control, after all, is with the host itself. The criticism of Shim6 was the basis for the design of Six/One.

4.2 Six/One: More Control for Network Operator

Six/One [4] is a host-based address mapping solution that gives edge network operators fine-grained control over which provider is used for traffic exchanged with the Internet. Six/One is similar to Shim6. Like Shim6, it is a multi-homing solution for IP version 6. Edge networks get address space assigned from each of their providers. Hosts configure an IP address from each address space, and they can use their IP addresses interchangeably without disrupting active communication sessions.

Unlike Shim6, Six/One enables the edge network to overrule the path decisions that hosts make, so as to exercise policy or to balance load. To handle such a re-routing decision by the network, Six/One introduces new functionality in routers on the border between the edge network and a provider. These routers rewrite the source address in egress packets so that the IP address always corresponds to the egress provider, even if the packet was re-routed. Hosts recognize the address change and adapt. Subsequent packets are then sent directly via the address that corresponds to the network-selected provider, without rewriting in routers.

Figure 6 illustrates Six/One's edge-network-initiated re-routing capability. If the host to the lower left in the figure initiates a communication via IP address 2000::def, that communication session by default goes via provider 2. The edge network may then decide to route the session via provider 1 instead. The peer

Improving the Scalability of Internet Routing

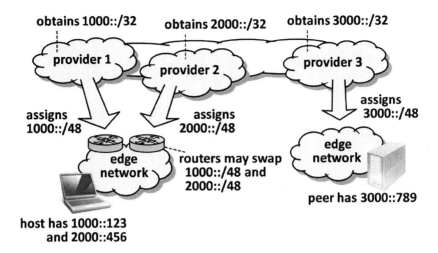

Fig. 6 Six/One

recognizes the re-routing first and switches to the provider 1 for subsequent packets that it sends to the host. The host recognizes the re-routing when it receives the next packet from the peer with the new destination address. The result is that the edge network has moved the packet exchange of the host and the peer from provider 1 to provider 2.

In order to maximally simplify the operation of a border router, hosts generate their IP addresses such that the IP addresses differ only in the prefix that identifies the provider. This is oftentimes the first 48 bits of an IPv6 address. So it is sufficient for a border router to rewrite only this prefix. This is a stateless operation; it can be done without host-specific or communication-session-specific knowledge.

Unfortunately, neither Shim6 nor Six/One eliminates renumbering in edge networks. Both of them rely on edge networks to use provider-assigned address space, and this address space changes as an edge network changes providers. To eliminate renumbering, network-based address mapping solutions like LISP and Six/One Router were designed.

4.3 LISP & Six/One Router Eliminate Renumbering

LISP [3] and Six/One Router are two of the most-discussed network-based address mapping solutions. The core idea is the same for both: IP addresses that are used inside edge networks are decoupled from the IP addresses based on which packets are routed across providers. The IP addresses used across providers can so be provider-assigned without creating a provider lock-in for edge networks, while the IP addresses used inside edge networks can be provider-independent without adversely affecting routing scalability. The IP addresses in each packet are then always mapped twice en route of the packet: At the sending side, a border router maps provider-independent IP addresses to provider-assigned IP addresses. At the receiving side, another border router maps the provider-assigned IP addresses back to corresponding provider-independent IP addresses. The mapping between provider-independent and provider-assigned IP addresses can be dynamic to support edge networks that are multi-homed.

Both LISP and Six/One Router follow these rules. They differ in how they achieve the mapping: LISP uses encapsulation on the sending side and decapsulation on the receiving side. That is, packets with provider-independent IP addresses are tunneled across providers inside packets with provider-assigned addresses. Six/One Router uses forward- and reverse-translation instead. This has two advantages: First, it avoids issues with packets exceeding the maximum supported size, because packets do not grow in size when leaving the Internet edge. Second, it simplifies backwards compatibility because most applications still function with packet exchanges in which only one of the edge networks supports the address mapping.

Nevertheless, all network-based address mapping solutions have a common problem: To work optimally for all applications, new infrastructure is required in both the sending and the receiving edge network. And this is a significant deployment hurdle, since edge networks that do not support address mapping are expected to exist for a long time. Proposals have been made to handle unilateral deployment of address mapping, but these are insufficient: Proxy address mapping has been proposed for LISP. But proxy mapping requires new infrastructure by itself. Six/One Router – as just observed – can deal with unilateral address mapping for a large class of applications. But since address translation is not transparent to applications, disruption of some applications cannot be ruled out.

5 Combining Advantages

The finding that host-based address mapping solutions fail to eliminate renumbering, and that network-based address mapping solutions are difficult to deploy due to new infrastructure requirements, suggests that address mapping in general is not the right solution approach. But perhaps there is a way to combine the advantages of both address mapping types. Such a combination would have to have the following properties:

- *Address mapping in hosts* — If a routing scalability solution were to do without new infrastructure, then it would have to be host-based. Network-based solutions, after all, require new infrastructure in both the sending and the receiving edge network, as has become clear above. On the other hand, the only infrastructure that host-based solutions may use is the DNS, and this already exists.
- *Provider-independent namespace in hosts* — If a routing scalability solution were to simplify renumbering, then it would have to incorporate a provider-independent namespace. Network-based solutions use a provider-independent namespace inside the entire edge network. This *eliminates* renumbering. Host-based solutions can maximally use a provider-independent namespace inside the host. This does not eliminate renumbering, because it does not remove the need to update an edge network's routing system upon a provider change. But it

simplifies renumbering because it removes the need to update hosts. The reduced cost of renumbering may make renumbering acceptable.

A solution that combines the advantages of host-based and network-based address mapping would consequently have to be host-based, and it would have to use a provider-independent namespace inside hosts. This approach has long been known as *identifier-locator separation*, where the provider-independent names form identifiers, and the provider-assigned IP addresses form locators. And indeed, identifier-locator separation enables multi-homing and simplifies renumbering: Multi-homing becomes possible because communication sessions are no longer bound to a particular IP address. The IP address can therefore change throughout the communication session without disruption. Renumbering becomes less problematic because hosts and applications no longer have to be updated when an edge network changes providers. They use identifiers that remain stable in such an event. The cost of renumbering therefore reduces, even though it is not completely eliminated.

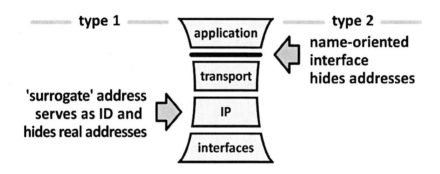

Fig. 7 Types of identifier-locator separation

Identifier-locator separation has been proposed in manifold variants, yet all of them can be grouped in two types. Figure 7 juxtaposes these two types:

- *Transparent identifier-locator separation* functions without application changes. It introduces a provider-independent identifier at IP layer, and uses this to hide from the application IP addresses that may be provider-assigned. The Host Identity Protocol is an instance of transparent identifier-locator separation. It uses a provider-independent "host identifier" at application layer in lieu of a real IP address. Mobile IP, too, implements transparent identifier-locator separation. It uses a provider-independent "home address" as an identifier, thus hiding the real IP addresses from applications.
- *Non-transparent identifier-locator separation* does require changes to applications. Existing variants introduce a new, name-based application-programming interface, which enables applications to initiate communication sessions directly with the peer's DNS name. The applications do not see an IP address.

Counter-intuitively, transparent identifier-locator separation is rarely deployed, whereas non-transparent identifier-locator separation enjoys increasing adoption. Name-based application-programming interfaces are used in several high-level programming languages and programming frameworks, such as Java, Web service frameworks, and peer-to-peer frameworks. And both new applications as well as new versions of existing applications rapidly adopt name-based application-programming interfaces. Why then does application transparency, a property meant to simplify deployment, turn out to be a deployment hurdle? The reason is three-fold:

- *Extra administration overhead* — Since applications today use IP addresses in identifying their peers, application transparency requires the identifiers in any identifier-locator separation solution to be syntactically indistinguishable from IP addresses. Such identifiers need to be newly introduced, as they do not exist in the classic Internet architecture, and with this introduction comes the need to map identifiers to the corresponding IP addresses. The mapping, in turn, requires infrastructure and, thus, human administration for unrestricted reachability of an identifier.
- *Risk of disrupting applications* — While application transparency avoids the need to port applications to a new interface of the operating system, it cannot assuredly prevent disruption of unmodified applications. Identifier-locator separation modifies the behavior of the operating system, and such modification may interfere with applications that rely on the original behavior. The subtlety and, for application developers, unpredictability by which this can happen makes such interference difficult to cope with.
- *Limited benefit for applications* — Application transparency, by definition, conceals from applications the service that an identifier-locator separation solution provides. Applications that rely on such service can hence be expected to implement identifier-locator separation themselves, thus making the same service in the operating system redundant.

These disadvantages are barriers for the adoption of transparent identifier-locator separation in operating systems and in actual use. Explicit name-based application-programming interfaces do not have these disadvantages. Application transparency therefore in fact creates a deployment hurdle, and it does not aid deployment as commonly believed.

6 Possible Way Forward: Name-Based Sockets

The foregoing analysis suggests that a name-based application-programming interface is the right approach to move forward with. But there is work to be done nonetheless: Existing name-based application programming interfaces do not support multi-homing because they do not incorporate coordination between a host and its peer. One needs to add multi-homing support, and hence one needs to enable hosts and peers to coordinate which IP addresses to use.

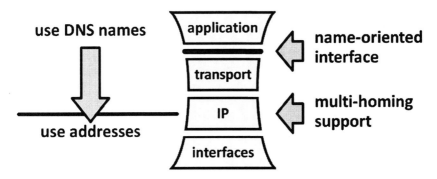

Fig. 8 Name-based sockets

The addition of multi-homing support to a name-based application-programming interface yields a name-based evolution of the sockets interface, *name-based sockets*, which is illustrated in figure 8. This takes the existing indirection between DNS names and IP addresses, which today is at application layer, and moves this down to the IP layer. Applications and transport protocols then bind their sessions to DNS names rather than to IP addresses, making the IP layer responsible for mapping the DNS names to the corresponding IP addresses. This mitigates the Internet routing scalability problem because:

- *It can simplify renumbering* — A systematic use of name-based sockets replaces IP address referrals in applications, scripts, and configuration files by DNS name referrals, and thus eliminates renumbering of hosts. Eliminating host renumbering, in turn, simplifies the overall renumbering procedure.
- *It enables multi-homing without new infrastructure* — Since communication sessions are bound to DNS names, IP addresses can change on the fly without disrupting the sessions. The only infrastructure that is needed for this is the DNS, and this already exists.

There are three more components that name-based sockets would need. In a nutshell, these are as follows:

- *Initial name exchange* — A method to exchange DNS names during the establishment of new communication sessions, so that a session can be identified by DNS names bilaterally instead of only on the side of the session initiator.
- *Backwards compatibility* in three regards: First, legacy applications must be able to communicate without using name-based sockets. Second, applications that do use name-based sockets must be able to communicate with legacy peers. And third, hosts should not be required to have a registered DNS name, even if they use name-based sockets.
- *Security* — Since communication sessions are now bound to DNS names instead of IP addresses, it is important that DNS names cannot be faked. This holds in particular for the DNS name of a session initiator, which is sent to the peer during session establishment. The peer should be able to establish the authenticity of the initiator's DNS name.

Neither of these components would be difficult to realize. An initial handshake can be realized by piggybacking the initiator's DNS name to regular packets at the beginning of a communication session. This is easy for transport protocols with inbuilt handshake such as TCP, because the transaction-oriented nature of a handshake makes it possible to piggyback the initiator's DNS name only to the first packet in the session; the second packet is then automatically an acknowledgment that the DNS name was received. For transport protocols without handshake – that is, for UDP –, the initiator's DNS name must be sent several times for reliability purposes, since it is not clear which of the packets are actually received. The responder should also return an explicit acknowledgment.

Backwards compatibility, too, can be provided: Legacy applications can continue to be supported by offering the classic IP-address-based sockets as an alternative to name-based sockets. Applications can then opt in to using name-based sockets. Legacy hosts as well as hosts without registered DNS name can be identified by synthetic DNS names, derived from an IP address during the establishment of a communication session. A synthetic DNS name derived from IP address A would then have the semantics: "the host which had IP address A at the time of session establishment." So the semantics of a synthesized DNS name would remain valid even if a host is multi-homing-capable and changes its IP address in the middle of a session.

Security can be accomplished through a forward DNS lookup by the peer: If the peer looks up the initiator's DNS name and finds that the retrieved IP addresses include the IP addresses that the initiator is using, then this is an indication for the peer that the DNS name received from the initiator is authentic. If unsecured DNS is used for this, the indication is weak, but perhaps sufficient in most cases. Where secure DNS is used, the indication is strong. Finally, several mechanisms exist, and can be re-used, to enable multi-homed hosts to securely re-route a communication session from one of their IP addresses to another.

An important question that comes to mind regarding name-based sockets is whether the new dependencies on the DNS are acceptable. Will the DNS cope with the new demand? Even though not every host will need a registered DNS name, a broad use of name-based sockets will inevitably increase the load on the DNS in terms of both, more lookups and more dynamic updates. There would be more lookups because a lookup would need to be made for potentially every communication session. There would be more dynamic updates by multi-homed, or possible mobile hosts who want to update the IP address at which they can currently be reached. Problems that may result from this are insufficient scalability and insufficient convergence. It was never tested how the DNS would behave if almost every host had a registered DNS name. It was also never tested how the DNS would behave if almost every host updated its IP address in the DNS upon a multi-homing event. And then there is the well known issue that many existing DNS implementations do not support low time-to-live values, which means that the convergence latency for dynamic DNS updates could be high. The fact that this is a bug, and not a feature, is irrelevant; what matters is the DNS implementations that are deployed.

Nevertheless, analytical results so far are promising. Several studies have been made and the results are available as papers. In particular, the new load would only affect the DNS servers at the bottom of the DNS hierarchy, and those could certainly be provisioned appropriately. After all, the work that a DNS server at the bottom of a hierarchy would have to do is conceptually less than that of a Mobile IP home agent, and Mobile IP home agent already exist. Support for low time-to-live values is achievable as well: First, it is a local problem. If a host wants to use name-based sockets, it is enough that the host's administrator or provider fixes its DNS servers. No one else needs to do something about the bug. Second, there is already a push to add security features to DNS servers. Support for low time-to-live values could be added alongside with the security features. And finally, fixing the DNS servers is not a prerequisite. Name-based sockets work also with legacy DNS servers; they just do not work optimally.

7 Conclusion

This chapter has shown that a silver bullet to fix the Internet routing scalability problem has so far not been found. In particular address mapping – a solution approach that had for a while been deemed very promising – has turned out to be insufficient, since its variants either do not simplify renumbering, or are too difficult to deploy. Identifier-locator separation could be a way forward. This would enable multi-homing in a more deployable manner. And by eliminating renumbering of hosts, identifier-locator separation would simplify the overall renumbering procedure.

There is, of course, still much work to be done. Calling complete the search for solution approaches of the Internet routing scalability problem would certainly be premature. Furthermore, many engineering details still need to be worked out. Test deployments, too, will have to be made. And not to talk about the deployment challenge.

So the current status is only a beginning. But the work continues, and hopefully other researchers will become interested as well. No one says that the work to be done will be easy. But hopefully this chapter has shown that research on Internet routing scalability is necessary, that the problem is solvable, and that the work will pay off — not only for the Internet of the future, but also for the Internet of today.

References

1. http://www.cidr-report.org (accessed April 2010)
2. Nordmark, E., Bagnulo: RFC 5533 - Shim6: Level 3 Multi-homing Shim Protocol for IPv6 (2009)
3. Farinacci, D., et al.: Draft-ietf-lisp-06.txt- Locator/ID Separation Protocol, LISP (2010)
4. Vogt, C.: Draft-vogt-rrg-six-one-02 - Six/One: A Solution for Routing and Addressing in IPv6 (2009)

Delivering Building Blocks for Internet of Services: Trust, Security, Privacy and Dependability

Aljosa Pasic

Atos Origin, Atos Research & Innovation, Albarracin 25, 28037 Madrid, Spain
aljosa.pasic@atosorigin.com

Abstract. One of the main pillars of European Future Internet strategy, so called Internet of Services, has been the central topic in European Technology Platform known as NESSI. The service centric view of Future Internet is also changing the way IT infrastructure and applications will be managed and delivered. In this context, trust, security, privacy and dependability considerations are playing major role when it comes to Internet of Service building blocks. While loosely coupled and globally distributed services are becoming norm for business process implementation and emerging business models, control processes and infrastructures are also expected to adapt to this environment. In addition, requirements for integrated compliance management and near real time reaction to policy violations are introducing new motivations and challenges for researchers. Managing assurance, security and trust for services, a NESSI strategic project, presents comprehensive solution to the problems posed by these challenges.

1 Introduction

The EU Research Framework Programme (FPs) have been the main financial tools through which the European Union supports research and development activities covering almost all scientific disciplines. FPs have been implemented since 1984 and the current FP is FP7, which has been operational since 1 January 2007 and will expire in 2013. Inside of FP7 there are 7 Specific Programmes under FP7, one of them being Cooperation. In the Cooperation Specific Programme of FP7 there are 10 Thematic Priorities and the ICT thematic priority (usually called ICT program), in its turn is subdivided into challenges. One of the most important new initiatives related to Challenge 1 of ICT program is Future Internet Initiative. The first phase of this public-private partnership was signature of so called Bled declaration, signed by more than 40 European research projects. Building upon the obligations of individual project contracts and the goals of the Strategic Agendas of the European Technology Platforms, this initiative started to work together through a European Future Internet Assembly in order to jointly design services and networking architecture for the Future Internet. The recent communication from the Commission to the European Parliament, the Council, the European Economic and Social Committee and the Committee of the Regions [1] is setting the stage for implementation of this ambitious objective. The other relevant documentation includes internal analysis of possible organizational models and their aspects in the implementation [2] and Brochure by the European Commission, Information Society and Media DG [3].

However, one of the main pillars of European Future Internet strategy, so called Internet of Services, has been the main topic for European Technology Platform known as NESSI (Networked European Software & Service Initiative) [4] . Inside of this initiative, an open framework, reference architecture and a number of strategic projects are promising to deliver some of the key building blocks for future Internet of Services. In this paper we will give a short overview of these initiatives with special focus on trust, security, privacy and dependability challenges and NESSI strategic project called MASTER.

1.1 The Story of NESSI

European Technology Platforms were proposed by the European Commission as an instrument to address innovation challenges in a coherent way and in domains that are strategic to Europe's economy. It in this context that NESSI launched in September 2005, at the initiative of 12 industrial organisations and one open source consortium that rapidly evolved into 22 partners and over 360 members. Today, there are over 30 other ETPs in areas ranging from manufacturing to intelligent textiles, as well as 6 JTIs – Joint Technology Initiatives. While many Joint Technology Initiatives emerged from ETPs, there is an essential difference: a JTI is set up as a legal framework, financed through a variable mix of industry, the European Commission and the member States. JTIs organise formal calls, similar to the research calls organised under the 7th Framework Programme by the European Commission but whose funding rules can differ. ETPs on the other side are coordinated by partners, and no public finances intervene in their creation or management.

When NESSI was launched at the end of 2005, the first few months were devoted to establishing clear governance and moving from the vision to the first Strategic Research Agenda. In 2007 NESSI produced not only additional iterations on the Strategic Research Agenda but also the way to achieve its vision: the NESSI Open Service Framework. Since 2008, NESSI focus among other things on building this framework through an open and world-wide process where **NEXOF-RA (NEXOF Reference Architecture)** project [5] is the first step that would enable the creation of service based ecosystems. NEXOF-RA main results will be the Reference Architecture for NEXOF, a proof of concept to validate this architecture and a roadmap for the adoption of NEXOF as a whole.

1.2 Future Internet of Services

To fully understand the positioning of NESSI in respect to Future Internet, the convergence view has to be taken as a starting point. Today, data and information are used pervasively in distributed networks and applications, while the frontier between objects and users is blurring. What is really linking all these future internet components is notion of services – users access the information they need in a specific context, using dedicated or generic terminals, but how the data is accessed, transformed and delivered, actually relies on services.

The strategic aspect of the Future Internet is therefore totally dependent on services, and this is why NEXOF underlines need to collaborate on international scale. In this holistic view, there is a clear identification of the framework whose aim is to deliver core building blocks, such as to protect privacy or to ensure interoperability.

2 Building Blocks for Internet of Services

Contributing to NEXOF is a process that has different channels, and is open on a world-wide basis. However, the main contributions to it come from so called NESSI strategic projects. Today, there are 6 NESSI Strategic Projects, with one of them, namely NEXOF-RA taking the role of coordinating the contribution process. The Strategic Projects involve 60 organisations with a research budget of about 120 M€, divided between industrial funds, European funds as well as national EU member states funds.

2.1 Closer Look at Challenges for Trust, Security, Privacy and Dependability in Future Internet

The service centric view (i.e. the notion that more IT will be delivered through the service lifecycle) is changing the way IT infrastructure and applications will be managed and delivered, and, as such, will be the context for end-to-end security considerations. The move towards services also increases the emphasis on relationships, negotiations, and agreements. This brings particular challenges for the area of trust, security, privacy and dependability, which are traditionally very difficult to manage and measure. It also forces the issue of accountability and liability. For all these reasons, a coherent framework for security actions and measures as well as management of interactions and dependability on other architecture elements, has to be prepared [6].

In Future Internet trust is no more simply assumed. The users of the internet are not a close knit group but span the entire globe. Many communication barriers have been broken and culture, language and distance are not as constraining as previous. Existing security mechanisms have to play a role, but their consideration as an add-on, as opposed to integral to the initial design have been widely recognized as prone of errors and is not satisfactory. New security mechanisms must be flexible and designed for change, evolution and adaptation in line with other Future Internet paradigms and able to resist unpredictable threats. In this ambit, security mechanisms must be designed to automatically configure and self-optimize themselves with respect to several dimensions, e.g. risk or context. The increasingly distributed, autonomous, open and heterogeneous nature of the current and future challenges demands for a coherent set of methodologies, techniques and tools to identify, assess, monitor and enforce system and services behaviour.

Another challenge is more related to preserving European societal values, and study architectural and system wide consequences, threats and risks related to these values. These might include the loss of privacy, transparency and accountability in communications and service provision chain, an open and fair operation and use of future internet that permits seamless cooperation and a competitive e-service market. The lack of accountability in today's Internet, for example, is demonstrated by the distributed denial-of-service attacks, spam, or phishing. At the IP layer all Internet traffic is almost anonymous, due to e.g. ease of source address spoofing and proliferation of network address translation. Many unwanted IP packets are sent by computers running programs unknown to their owners.

More than ever, the Future Internet will be characterised by distributed storage and processing of data. This means that network nodes are not only used for data transmission and terminals are not only used for application processing. The former become hosts for data and application services while the latter participate in an ad hoc, more or less controlled manner, in data transmission through the network. This is a kind of convergence that transcends every layer of the network and it goes hand in hand with the convergence between mobile communication networks and the core Internet.

There is also a number of security challenges that will result from the evolution of software systems in the future internet, for example internet-centric operating systems, semantic support (context recognition, automated adaptation), availability of virtualisation resources, complexity modelling and management and execution environments. Finally, while availability is well understood in traditional systems, new research is required to face the challenges of heterogeneity, massive scale, and mobility in the internet of services.

The NESSI trust, security and dependability (TSD) model relies on complete requirements that include business, technical, legal, regulatory, and societal requirements. Building blocks and research topics that have been identified have focus on (a short selection):

- Security services: cross-cutting basic and specialised services
- Knowledge representation: TSD properties representation, policy languages, etc
- Embedded Intelligence: Rules, mechanisms etc for negotiation, autonomic capabilities etc
- Dynamic assurance: metrics, trustworthiness...
- Securing Services: process-based tools and methodologies
- Secure Service Management: accountability, monitoring, control, compliance management...

Several FP7 projects have looked at one or more of these building blocks. Formal logic models have been developed while other aspects of trust – sociological, psychological, legal, ethical, economic, etc. – have also been studied. Some projects are in particular looking at architectural principles for a secure virtualized platform or secure service compositions. When it comes to international research

Delivering Building Blocks for Internet of Services

collaboration, frequently stated issues are knowledge representation or metrics. In the following subchapter we will present NESSI strategic project which is covering some of the above named challenges and blocks.

2.2 Managing Assurance, Security and Trust for Services: Master Project

Move towards the new ICT paradigms (such as Service Oriented Architectures (SOA) or cloud computing) and business models (e.g. outsourcing) is sometimes perceived as "losing control" over some business process steps, and therefore it is potentially bringing new and unknown risks, primarily related to policy enforcement and regulatory compliance. The risk relationships and decision making becomes more complex and, in addition, there is a reduced time to react. Compliance with regulations, however, mandate enterprises to provide enough evidence [7] to auditors so that those auditors can judge whether regulations have been obeyed or violated and in the case of outsourced IT services or processes some regulations still hold an enterprise (or even the CEO of an enterprise) liable, even if an outsourcing provider violated a regulation.

Therefore static compliance and evidence collection for audit process proves not to be sufficient [8]. While we recognise that the proper choice, design and deployment of security controls remains an essential part of IT security consulting practise related to the regulatory compliance, this chapter describes a different approach and set of mechanisms to enforce compliance in Internet of Services.

Master (which stands for Managing Assurance, Security and Trust for Services) is a European Commission project inside the Seventh Framework Programme for Research and Technological Development. In order to provide methodologies and tools that facilitate monitoring, enforcing, and auditing compliance in a highly dynamic service- oriented architecture, MASTER project introduced a number of indicators that could be used to measure and assure trust, security, privacy and dependability in Internet of Services.

Compliance has been investigated in several contexts. It is also frequently treated by different ICT stakeholders: auditors, risk managers, security officers etc. Related tools and technologies span from compliance modeling tools, through paradigms such as Business Activity Monitoring (BAM) or Internal Control Systems (ICS), to a wider areas of Governance, Risk and Compliance (GRC) and Enterprise Risk Management (ERM). On the market there are already some solutions for centralizing collection of compliance data, including data from the high-level business intelligence (BI) tools down to a lower-level IT management tools like service management or configuration management products.

Compliance in MASTER is treated through two main streams of research: while compliance engineering uses risk-driven goal decomposition [9] in order to interpret high level compliance objectives (often expressed in natural language) and maps them into the "observable" indicators, the second research stream takes inputs from signaling and monitoring infrastructure and tries to achieve near real time reactive compliance enforcement. Architectural components are based on techniques and tools for analyzing events produced by the MASTER infrastructure in order to perform detection of policy violations, analyse causes of policy violations, derive and predict models for violations, and provide compliance reporting.

One of the main problems we have to deal with is the translation of non-trivial business goals, compliance requirements and organisational policy aspects into technical controls. While most of the translation may be straightforward, it is frequently easy to insert errors that may not be obvious even on review (e.g. transposed entries in access control lists). The complexity of all such technical translations is likely to overwhelm an individual or even a group that is attempting to validate the conformance of the technical translation of the natural language security policy and detect inconsistencies. Even if a clear and unambiguous translation mechanism existed, components such as operating systems, but also middleware layers, usually can implement only simple security controls that are frequently insufficient to translate exactly the higher level compliance policy statements and requirements. We call this "semantic" compliance gap or **"expressiveness" challenge**. In matter of fact, when it comes to (semi)-automated control mechanisms, the most important aspects are related to the establishment and mapping of policies that have different level of abstraction, which is closely related to **"granularity" challenge and issues of cost, complexity, scalability and performance**. Right metrics and indicators that are basis for triggering reactive components, as well as the overall "cost" model with a dynamic weight assignation for the each process, according to its relevance and impact in a integral model. Financial processes, for example, that is subject to Sarbanes Oxley (SOX) Act [10] compliance needs to have more strict compliance control mechanisms than some other processes (e.g. approval of travels). Section 302 of the SOX Act, for example, mandates a set of internal procedures designed to ensure accurate financial disclosure and impact of non-compliance can be huge. According to it, the signing officers must certify that they "have designed such internal controls to ensure that material information relating to the company and its consolidated subsidiaries is made known to such officers by others within those entities, particularly during the period in which the periodic reports are being prepared". Translating this statement into an operation security policy that can be monitored and whose violation creates non-compliance evidence, is an evident challenge.

The MASTER enforcement infrastructure provides the means for making sure that the constraints formulated in MASTER operational security policies are satisfied. While the signaling and monitoring policies specify what evidence must be collected to assess the runtime parameters of the MASTER infrastructure or to enact its constraints, the enforcement policies specify what actions must be performed when an event of interest happens. Based on incoming notifications, the enforcement components perform the necessary actions, according to the enforcement policies. These actions can be either preventive (e.g. call inhibition, call modification) or reactive (e.g. cutting access, reconfiguration, undoing certain actions).

Generally, the format of a policy is split in three distinct parts: an event being triggered, a condition accompanying the event, and an action to be performed in that particular situation. Following this Event-Condition-Action (ECA) model, the enforcement framework is concerned with performing the action, and the monitoring framework with the event firing for the enforcement components once the condition specified in the policy is evaluated to be true. For the current implementation, we made an assumption that the monitoring has the capability to perform condition evaluation. In reality and in relation to the outsourcing scenario, the enforcement might not be within the same trust domain as the monitor component. That is, the monitoring infrastructure may have different trust level assigned (e.g. according to service provider reputation) for each part of outsourced business process, so that some enforcement components, placed at service provider with higher trust level, could perform its own condition evaluation. This decoupling of monitoring and enforcement infrastructures corresponds partially to a solution of **"end-to-end"** trust, security, privacy and dependability challenge. It is also bringing unprecedented control **flexibility as well as adaptability** to dynamic business scenarios (e.g. virtual organizations). In matter of fact, analogical to multiple trust domain scenarios in business process outsourcing, which is considered by enterprise governance and risk levels, we have multiple trustworthiness levels for distributed software components deployed as services, which have to be considered at information security governance and risk levels. The final result is interplay of governance and risk elements where scale of trust has to be iteratively fine-tuned and where concept of **dynamic trusts and linkage between trustworthiness and trust** come forward. For this purpose, the interaction between the monitoring and the enforcement infrastructures occurs in both directions. First, the enforcement components subscribe to evidence issued by the monitoring. For this reason, anytime the monitoring triggers an event notification (accompanied or not by a condition evaluated by the monitor), the enforcement components are alerted. Second, because the enforcement modifies the application runtime, its actions and decisions must be recorded as events; as a consequence, the enforcement framework will have to issue event notifications that the monitoring component will capture.

As mentioned before, there are two types of enforcement components: preventive components and reaction components. Both of them subscribe to events produced by the monitoring and signaling components, as specified by the enforcement policies. These enforcement policies are supplied to the policy repository and distributed to both kinds of enforcement components. Since the business context dictates the semantics of the actions specified in the enforcement policies, the enforcement components need to interact with the domain ontology in order to infer the meaning of the actions they have to perform.

Architecturally, the prevention and reaction components are similar (see component diagram Fig. 1). The main difference is that reaction component looks for policy violations, which in their turn are combinations of conditions and events (already captured and evaluated by the monitoring) that have already occurred. The most important subcomponents of an enforcement component are the policy interpreter and the rule engine. The former is able to consult an external ontology service and derive the logic behind every policy fed to the enforcement framework. Once the enforcement system knows what it has to do, it subscribes to the right events; once it starts getting some of these events, The rule engine maps them to the right policies and then performs an enforcement action as stated in the

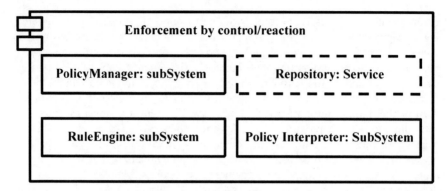

Fig. 1 MASTER Enforcement component.

enforcement policy. Finally, it publishes the action it performed for the monitoring component to observe. If we use concepts introduced in the previous sections, and define policy violation through use of indicators, we can also say that reactive enforcement will be triggered when KSI/KAI threshold has been reached.

Finally, we should say that MASTER reactive enforcement first implementation is at two SOA levels of abstraction: at the BPEL level and at the ESB level. BPEL is concerned with the orchestration of business actions, so enforcement can take the form of temporarily blocking, denying, modifying or even inserting a business activity from / into the original business flow. Potential triggers could be any high level event ranging from a variable assignment to a state change in a business activity.

While BPEL is concerned with process logic ("what are these services talking about? What is this service interaction actually doing and how?"), ESB is concerned with delivering the conversation to the involved parties. BPEL cannot manage message mediation because it does not consider what happens to a message once it is released and before it is received. For this reason, enforcement capabilities at the ESB level make up for what BPEL enforcement cannot achieve at the business level, in terms of SOA requirements. For example, BPEL policies cannot deal with constraints of the type "any event emitted by service A destined to service B must pass through service C first". This rule can be easily fed to the ESB engine and would be immediately enacted.

3 Conclusions

NESSI is defined in the context of a holistic approach to an ecosystem in which all the parties involved coexist and which can develop into a new economic model. This holistic model embraces the whole service area and foresees NESSI as a key element in the EU, but also in the global economy.

A growing emphasis on international R&D collaboration is a measure of the development of a global knowledge-intensive economy. International cooperation agreements enables but does not guarantee successful cooperation, and this does not automatically lead to innovation, the introduction of new goods, services, or business processes in the marketplace. Differences in national systems of European R&D funding schemes (for example, support for collaborative projects) may make one country more effective than another in international cooperation.

When it comes to TSD research and building blocks described in this paper, we can say that TSD research is currently highly diversified and generally focuses on short to mid-term returns. One of the advantages that future Internet of Services brings is actual decoupling of process steps that are now implemented as separated web services. In the future this will enable a large number of service constellations and an explosion of global service chains based on the emerging business models [11]. In its turn we will be forced to decouple and globally distribute also other processes that actually do not create business value (such as control processes). Monitoring security relevant events at global scale brings many new challenges, ranging from scalability to the legal issues. This has been exemplified with situations and solutions, such as the one proposed in MASTER, where control activities are wrapped around outsourced business process steps, and where evidence collection and condition evaluation can happen in different countries.

The Future Internet research collaboration on international scale is becoming compulsory, in order to ensure that end-to-end global service-based scenarios are consistently and coherently implemented. This collaboration is also positive for the optimization of resources, especially in time like these when research resources are also affected by crisis.

Over the past two decades, R&D has principally been performed and funded in North America, Europe, and Asia. However, R&D expenditures are estimated to have risen rapidly in selected Asian and Latin American economies and elsewhere so these countries could and should be incorporated in the Future Internet joint research activities.

References

1. http://ec.europa.eu/information_society/activities/foi/library/ficommunication_en.pdf (accessed December 2009)
2. http://www.future-internet.eu/fileadmin/documents/reports/Final_Report_Model_PPP__270409_.pdf (accessed December 2009)
3. ftp://ftp.cordis.europa.eu/pub/fp7/ict/docs/ch1-g848-280-future-internet_en.pdf (accessed December 2009)
4. http://www.nessi-europe.com/Nessi/ (accessed December 2009)
5. http://www.nexof-ra.eu/ (accessed December 2009)
6. Pasic, A., Serrano, D., Soria, P., Clarke, J., Carvalho, P., Maña, A.: Security and Dependability in the Evolving Service-Centric Architectures. In: At Your service. MIT Press, Cambridge (2009)
7. Miseldine, P., Flegel, U., Schaad, A.: Supporting Evidence – Based Compliance Evaluation for Partial Business Process Outsourcing Scenarios. In: The 1st International Workshop on Requirements Engineering and Law, Barcelona, Spain, September 9 (2008)

8. Basin, D., Burri, S., Karjoth, G.: Dynamic Enforcement of Abstract Separation of Duty Constraints. In: Backes, M., Ning, P. (eds.) ESORICS 2009. LNCS, vol. 5789, pp. 250–267. Springer, Heidelberg (2009)
9. Refsdal, A., Stølen, K.: Employing Key Indicators to Provide a Dynamic Risk Picture with a Notion of Confidence. In: IFIPTM proceedings, West Lafayette (USA), June 15-19 (2009)
10. http://en.wikipedia.org/wiki/Sarbanes%E2%80%93Oxley_Act (accessed December 2009)
11. Schleicher, D., Anstett, T., Leymann, F., Mietzner, R.: Maintaining Compliance in Customizable Process Models. In: 17th International Conference on Cooperative Information Systems (CoopIS 2009), Vilamoura, Algavre, Portugal, November 4-6 (2009)

ITU Focus Group on Future Networks

Nilo Pasquali and Abraão B. Silva

Agência Nacional de Telecomunicações (ANATEL), Brasília, Brazil
{nilo,asilva}@anatel.gov.br

Abstract. Next Generation Networks (NGN) is envisaged to "substitute" all telecommunication network infrastructure for a packet IP-based concept in the near future. However, is this NGN infrastructure capable to support all future applications, especially those that are highly dependent on quality, reliability and speed? To evaluate this, the standardization sector of the International Telecommunications Union (ITU-T) has started studies related to the Future of Networks, that is, networks beyond NGN.

1 Background

1.1 Anatel's Role, Why Are We Talking about This

Anatel is the National Telecommunications Agency in Brazil, responsible for the regulation of the telecommunication market. It was created on the period that Brazil stared the privatization of the sector, around 1997. Its main responsibilities are related to the implementation of national policies related to telecommunications, give authorizations for service providers, administer the radiofrequency spectrum and satellite orbits, and supervise the overall functioning of the Brazilian telecommunications market.

Under Anatel's responsibilities is also the task to represent the Brazilian administration in all international forums related to telecommunications [5]. It is no surprise that, from an Administration point of view, the ITU is the main international telecommunication body, so, Brazil takes special interest on its activities. In order to fulfill its responsibilities as representative of the Brazilian Administration, Anatel has a specific structure under its internal organization form by four Commissions (Brazilian Communication Commissions - CBCs) [6], each responsible for specific aspects of the telecommunication environment, they are: i) Governance and International Regimes; ii) Radiocommunications; iii) Telecommunication Standardization; and iv) Telecommunication Development.

Under the third Commission, namely CBC 3, is the responsibility for all the standardization work conducted by the standardization sector of the International Telecommunication Union (ITU). Therefore, in the scope of Future Networks, and the Focus Group conducting this activity, which will be better explained in the following chapter, the work is under CBC 3 responsibility. The main objective of this paper is to present the work that has just begun in the ITU on the matter, and invite interested parties to participate.

1.2 ITU Standardization Sector and the Creation of the Focus Group

The World Telecommunication Standardization Assembly (WTSA) takes place every four years in order to review working methods, approval process, work program and structure of the standardization sector Study Groups and the overall management of ITU-T. It undertakes studies, make regulations, adopt resolutions, formulate recommendations and opinions, and collect and publish information concerning telecommunication matters [1].

WTSA-08 took place in Johannesburg, South Africa, from the 21st to the 30th October 2008. Among all the work carried out during this period, Resolution 2 (Johannesburg) [3] is of special interest, since it establishes the responsibilities and mandates of the Study Groups of the Standardization Sector. The Fig.1 bellow gives a general idea of the structure of ITU-T´s work.

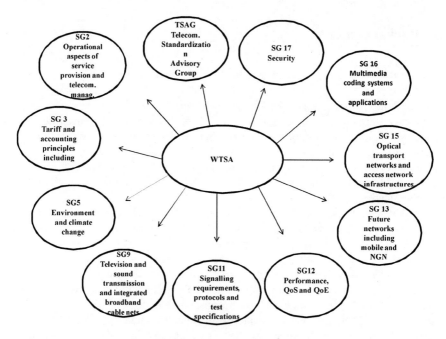

Fig. 1 ITU-T work's structure

Annex A to Resolution 2 points to the general area of study for each Study Group in ITU-T presented in the figure above, and states the following for Study Group 13, entitled "Future Networks including mobile and NGN":

"Responsible for studies relating to the requirements, architecture, evolution and convergence of future networks. Also includes NGN project management coordination across study groups and release planning, implementation scenarios and deployment models, network and service capabilities, interoperability, impact of IPv6, NGN mobility and network convergence, public data network aspects and network aspects of IdM. Responsible for studies relating to network aspects of mobile telecommunication networks, including International Mobile Telecommunications (IMT), wireless Internet, convergence of mobile and fixed networks, mobility management, mobile multimedia network functions, internetworking, interoperability and enhancements to existing ITU-T Recommendations on IMT."

Prior to WTSA-08, Study Group 13 had the responsibility for NGN related studies and Study Group 19 was responsible for mobile networks, including Fixed to Mobile Convergence (FMC). The revised Resolution 2 (Johannesburg) combined these areas of responsibility and expanded the work to include Future Networks.

The possible limitations to current IP-based network architecture for providing support to futuristic applications have being studied for some time, and many projects around the world are dealing with this matter such as IRTF (International), GENI/FIND (in the United States of America), FP7/FIRE (in the European Union), CNGI (in China), AKARI/NwGN (in Japan), FIF (in South Korea), ARCMIP (in Brazil), among others.

Question Q.21/13, under the purview of Study Group 13, was then assigned to study matters related to Future Networks. As stated on the motivation to this question, the Future Network is envisaged to be an evolution of the current IP-based network architecture, capable of providing futuristic functionalities that are unsupported by current technologies, or that their implementation is prohibitive under the current paradigm [2]. Essentially, there are concerns related to many aspects of IP based networks, such as scalability/ubiquity, security/robustness, mobility, heterogeneity, Quality of Service (QoS), re-configurability, context-awareness, manageability, data-centric, network virtualization, economics, among others. At first, backward compatibility with existing networks and systems could be understood as unnecessary, however, this idea has been changing as discussions progress.

The study items, allocated to question Q.21/13, as approved in WTSA-08, are listed as follows:

- Problem statements on current IP-based network architecture (which should consider the current vision of future services and applications)
- Design goals and general requirements for the Future Network (scalability, transparency, multi-homing, traffic engineering, mobility, security, robustness, re-configurability, context-awareness, manageability, heterogeneity, data-centric, and economics)
- Gap analysis between existing standards and/or proposals for next generation networks and the design goals and requirements for the Future Network (FN)

- Study of Meta architectures (network virtualization, cross-layer communications) and architectural framework for the Future Network (e.g. wireless, advanced photonics, embedded computing, intermittent network/DTN, vehicular/airborne network, programmable and cognitive radios, network virtualization, overlay service control)
- Study on how to incorporate new communication and service technologies into Future Networks, such as wireless edge network (e.g. mesh/sensor, ad-hoc, network movement) or optical backbone network (e.g. optical switch or router).
- Identify functions and capabilities necessary to support new services for the Future Network (e.g. user-centric, context-aware, user's preferences considering, proactive users provisioning, seamless services, QoS)
- What enhancements to existing Recommendations are required to provide energy savings directly or indirectly in Information and Communication Technologies (ICTs) or in other industries? What enhancements to developing or new Recommendations are required to provide such energy savings?

Therefore, it is clear that there is an extensive list of items to be studied during the four-year cycle (2009-2012) that just begun. To start this work, at the first Study Group 13 meeting, held in Geneva/Switzerland, from 12 to 23 of January 2009, question Q.21/13 met and proposed the establishment of a Focus Group on Future Networks (FGFN), since it allows for interested parties that are not members of the ITU to join the technical work, such as other standardization organizations, experts, individuals, academia, etc.

Additionally, at this first meeting of question Q.21/13 a preliminary workplan and general timeline was produced. These first ideas contemplated the development of a vision and service scenarios for Future Networks and two main Recommendations: a high-level requirements document and a framework document. All these should be concluded by the first quarter of 2012, and some work would overlap with the activities of the Focus Group, as it is the main input for the work of the question.

An overall idea of these two Recommendations can be summarized in the Table 1, using the names as attributed by question Q.21/13.

Table 1 ITU-T recommendations scope and objetive

Name	Scope and Objective
Y.FNvision	Definitions, Values, Visions, General Concepts and Requirements for Future Networks. General idea: "a network that is capable of providing revolutionary services, capabilities and facilities that are hard to be provided by current network technologies".
Y.FNvirt	Requirements for network virtualization, as required for Future Networks. This Recommendation will also investigate requirements, scenarios and procedures for network virtualization.

2 ITU´s Focus Group on Future Networks

The Focus Group on Future Networks was proposed and agreed upon at the first meeting of Study Group 13 for the 2009-1012 study period, with the scope to, in collaboration with worldwide Future Network communities, collect and identify visions of such networks, based on new technologies; assess the interactions between these networks and new services; familiarize ITU-T and standardization communities with emerging attributes of Future Networks; and encourage collaboration [4].

The first meeting of ITU´s Focus Group on Future Networks was held in Lulea, Sweden, from the 29th of June to the 3rd of July 2009, having one joint with the European initiative FIRE. The first step on this meeting was to agree on the working methods and start the studies related to initiatives around the world on the issue of Future Networks. A preliminary framework of existing activities in China, Europe, Korea and Japan were presented and a repository was created to gather all possible activities. At total, it was presented eleven project descriptions related to Future Networks.

Two main deliverables were identified, as a result of the initial work from question Q.21/13, mainly: i) Future Networks: vision, concept and requirements; and ii) Framework of network virtualization. The discussions on timeframe for prototyping and phased deploying of Future Networks were an issue, and general agreement understood that somewhere between 2015 and 2020 should be a rough estimate.

The general timeline and tasks of the work of the FG-FN were identified and agreed. For this first meeting the main tasks included the setting up of the Focus Group and its working methods and begin the review of ongoing activities/initiatives on Future Networks. The following meeting, to be held in November/2009 in the USA, in collocation with project GENI, should continue the main tasks started at the previous meeting and carry on with setting up external collaboration channels, establish the descriptions of existing activities and identify/describe benefits and visions for Future Networks. The following couple of meetings, in 2010, should then identify and describe attributes of these networks, build a vocabulary and draft action plans for further work.

All work related to ITU-T´s Focus Group on Future Networks is available online at http://www.itu.int/ITU-T/focusgroups/fn/index.html.

3 Conclusion

The Future Networks study is a key issue in order to ensure the continuum of the expansion and innovation on the Telecommunication sector. Even though many solutions have been found, and others are under study, to make current networks capable to support some of the envisioned applications, a bottleneck will be reached at some point. This "limitation" has to be dealt with and initiatives around the world have already started to deal with this.

The ITU-T, as one of the major standardization bodies in the world has a big responsibility on this area, so is no surprise that started it studies on this matter as well. Recognizing the need of participation from experts on the area, the ITU-T established the Focus Group on Future Networks, to ensure that it would be able to gather the most expertise in the issue, in an environment much easier to participate and collaborate.

Brazil is following closely all this work, since it has its own activities on Future Networks and will continue to participate long after the conclusion of the Focus Group work.

Acknowledgments. This work is supported by the Agência Nacional de Telecomunicações (ANATEL), Brazil.

References

1. Constitution of the International Telecommunication Union, http://www.itu.int
2. ITU-T Contribution 1 to Study Group 13 from WTSA-08: Questions assigned to Study Group 13 (Future networks including mobile and NGN) for the study period 2009-2012 (December 2008)
3. ITU-T Resolution 2 (Johhanesburg): ITU-T study group responsibility and mandates
4. ITU-T Temporary Document 3 of WP 5/13. Terms of Reference of the Focus Group on Future Networks (FG-FN)
5. Law 9.472/1997. General Telecommunications Law (LGT). Brazil
6. Resolution n° 502, of 18/04/2008. Alters the organization structures of the Brazilian Communication Commissions

Key Issues on Future Internet

Tereza Cristina Melo de Brito Carvalho, Charles Christian Miers,
Cristina Klippel Dominicini, and Fernando Frota Redígolo

Laboratory of Computer Architecture and Networks (LARC),
Escola Politécnica - University of São Paulo (EPUSP), São Paulo, Brazil
{carvalho,cmiers,cdominic,fernando}@larc.usp.br

Abstract. The Internet has changed the way people live and interact with others. Everyday, an ever-increasing number of activities can be conducted through a myriad of Internet-connected devices. It is possible to foresee a Future Internet environment centered on users, where a variety of devices benefit from ubiquitous connectivity to augment every aspect of people's life. An incredible amount of applications will be created in a user-driven approach to provide data and services in a user centric way. These applications and services could be sold and used by other users to create their own new applications producing new information and services in an amazing speed. This scenario has far-reaching technological, economical, social and political implications. This chapter presents the main trends and challenges to be faced towards a more ubiquitous, interactive and user-centered Future Internet.

1 Introduction

The Internet has changed the way people interact among themselves, purchase goods, plan trips and vacation, handle their finance life, educate, learn, and as they live.

Nowadays people making use of different kind of devices and are always connected. Every day we live in even more ubiquitous environment, surrounded by mobile and fixed smart devices that keep us connected to our family, our friends, our work, our house, the worldwide news and help us to make decisions, perform our daily activities, share our data, develop collaborative tasks and improve our productivity.

In this scenario, new technologies have been developed considering new operational requirements, such as mobility, size, transmission bandwidth, storage and processing capacity and cost. New types of applications and human interfaces have been envisioned considering people with different abilities, backgrounds, cultures and needs that make use of different devices as well diverse communication mediums.

More than technological and economical issues, social and environmental issues have to be taken into account. In general terms, sustainability has become an increasingly important factor when we are handling technologies and processes that can cause positive and negative impact on people and nature. In addition, government regulations regarding resources and information usage, and the communication service provision also play an important role.

This paper presents the main technological trends and challenges to be faced towards a more ubiquitous, interactive and user-centered Future Internet. It is organized in four additional sections besides the current one. Section 2 explains the trends related to the Internet of the Future and provides a big picture of its scenario. Then, an overview of the technological key technologies and concepts of this new Internet generation is presented in Section 3, including Web 3.0, cloud computing, Internet of Things (IoT) and 3D Internet. Section 4 presents some technological issues and challenges to be faced for enabling this new generation of Internet while Section 5 brings some discussion concerning business models, human values and sustainability. Finally, the last section contains some considerations on Future Internet based on the issues previously presented.

2 Trends towards the Future Internet

According to (Hourcade et al. 2009), the main trends related to the Future Internet, its usage, architecture and infrastructure are:

- **We are always connected, making use of different devices in different places.** A very simple example is the access to our e-mail and contact list from our notebook in our office or at home, from our cell phone when we are transit (e.g., in a queue at the airport) or from our car system. In these cases we are deploying different types of terminals, different types of network (wireless or wired network in the office and at home, cell phone network and radio network in the car) and different service providers for accessing the same content.
- **Everyday we use more and more online-based services.** In the case of Brazil, it is especially true for home-banking services, however other services have become popular such as games (3D- Video Game), TV, digital cinema, on-demand video (e.g., HDV), newspaper and magazine subscription, e-mall shopping, distance learning and presence (social) networking. These services have different requirements in terms of display, computational and networking capacity and security.
- **The Internet has evolved rapidly from information sharing to collaborative production.** The Internet designed decades ago allowed different sites to share information in B2B, B2C, C2C relations. Since then, users have been empowered with different tools and low cost devices (e.g., web cameras), which allow them to create their own contents and personalize their products and services. Empowered users realized that instead of producing only their own content they could have collaborative production. This phenomenon has changed commercial, work and social relations.
- **Everyday the number of smart objects that surround us increases.** These objects collect and process data that are used later on for providing information and taking decisions. This is possible due to their embedded sensors, networking

and computing capacities. In this context, known as Internet of Things, we can think about different applications. One very interesting and useful is related to a "plant" that communicates with its owner asking for water or even vitamin. Implementing this depends not only on embedded humidity sensors (capable of detecting how dry is the ground) but also on communication and computing resources. When it is detected that the ground is too dry, an alarm will be generated to the plant's owner asking for water.

- **Personal Information is spread on different systems and networks.** Each user has different devices that interact with himself/herself and other users simultaneously. Currently, it is more common to have each device collect information to be only used in its internal software. However, in future it will be required that several user devices collect information about one or several users (location, time, behavior, etc) to be deployed in a set of systems configured for providing data to specific user applications. These applications can be configured previously or on-demand.

Fig. 1 presents an example of a scenario based on these Future Internet trends.

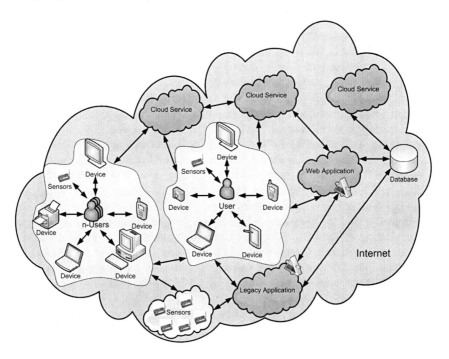

Fig. 1 Example of Future Internet scenario

In such cases, the key features are:

- **Shared content among different types of devices:** The content has to be synchronized among different systems (e.g., e-mail system) and adapted to devices with different characteristics, such as screen size, storage and processing capacity.

- **Context sensitivity and adaptation:** The devices can be configured to be context sensitive to the settings and privacy policies. For instance, as soon we enter in the airplane the cell phone switches off or when a meeting begins, it switches to vibration mode.
- **Ubiquitous user interface:** It is important to have a ubiquitous navigation and service user interface that is common across a heterogeneous multitude of digital media devices. It is especially important for the disabled and elderly people (Marcus 2003).
- **User-Generated Services and Content:** Technologies supporting user-generated content and services have been developed quickly, allowing users to combine existing content and services easily to provide novel integrated formats and services (Schonwalder 2009).
- **Mobility:** In the future, mobility is a feature of the whole system, including the mobility of end-points and whole edge networks (Gluhak et al. 2009). It requires management features such as the tracking of the set of active devices for a user and data routing making use of the optimal devices (Schonwalder 2009).
- **Quality of Service Provision:** Different types of applications and contents have different QoS (Quality of Service) requirements to be provided in different types of Internet Access. The parameters of QoS can include high-transmission rate, low delay, low delay variation and low data-error transmission. For instance, HDV (High-Definition Video) transmissions and 3D-Displays have very strict QoS requirements, such as low latency broadband connections.
- **Quality of Experience by the service user:** It corresponds to the overall value of the service provided from the user's perspective. The concept of QoE in engineering is also known as Perceived Quality of Service (PQoS). Moreover the PQoS evaluation gives the service provider and network operator the capability to minimize the storage and network resources by allocating only the resources that are sufficient to maintain a specific level of user satisfaction.
- **Standard Intercommunication:** Considering an environment of heterogeneous devices with different communications, processing and storage capacities as well different types of interfaces, heterogeneous communication infrastructure, including from sensor to optical networks with different operation parameters, different service providers, being used by multiple value-added applications and services with different QoS and security requirements, it is important to have a platform that makes possible a standard access to the basic common services such as communication and security services.
- **Metadata Identification:** Information can be persisted in different ways, including text, picture, sound or video. Furthermore, information generated by the real world will make sense when put into the right context. It is thus important to tag this information with metadata (e.g., geographic location and quality of information) for further processing (Gluhak et al. 2009). It requires new mechanisms to manipulate multimodal information under different criteria across a multitude of distributed hosts.
- **Easy Services customization:** The current Internet provides mass services with little personalization, but, in the future, services will be heavily personalized, either user-oriented or group-oriented (Schonwalder 2009). Personalization

brings additional value to the market, increasing the product appeal and customer reach. Thus, the simplification of the customization process, for instance, making use of person-machine interfaces and the quality of the customization experience become critical factors.

- **Dynamic Pricing and Accountability techniques:** As customers will be able to customize their services and products, their prices will vary from one customer to another. When you design your own running shoes or backpack, normally there are a limited number of design options that become easier to determine their price. On the other hand, if you want to buy a car insurance based on your driving behavior (e.g., car speed, type of trips, parking places), it requires a much more sophisticated dynamic pricing and accountability techniques.
- **Trust, Privacy, Security and Identity services:** Provided services rely on distributed systems and an heterogeneous infrastructure involving several parties. In such environment, identity management will be critical for managing trust, safeguarding privacy and supporting security services and policies. Concepts related to dynamic trust and security based on peer entities behavior and history have to be considered (Schweitzer et al. 2006). Privacy and Security will become increasingly important for Internet users as they want to have secure, trusted, and ubiquitous access to services in the ambient environment while being constantly confronted with sensors, cameras, mobile phones and networking devices (Schonwalder 2009).
- **Scalability:** The current Internet routing and addressing system is hitting its limits as more manual configuration is needed to avoid cascading problems due to overload, accidental misconfiguration, or attacks (Stuckmann and Zimmermann et al. 2009). Although the IPv6 standard allows expanding the address pool, the vast majority of software and hardware still use the IPv4 technology. As the Internet will have billions of connected users and objects in the future, one of the biggest challenges is to deal with addressing and routing in a scalable manner.
- **Service Paradigm:** Although a lot of computing and storage applications are still executed locally on end-user devices, a service-oriented Internet would allow access to complex physical computing resources, data, or software functionality in the form of services (Stuckmann and Zimmermann et al. 2009). For instance, Cloud computing has recently emerged from this movement as a mean to provide computational resources (e.g., virtual execution and storage) as Internet services (Schubert et al. 2009).

The key features listed above show that is necessary to create new systems and solutions fostering the Internet evolution. However, these new systems should coexist with the current ones, demanding technologies and solutions for the interoperability between current applications, services and networks with the newer ones.

3 Technological Concepts in the Future Internet

Worldwide, computing systems are moving quickly from local and homogeneous systems to distributed, heterogeneous, and collaborative ones. According to this reality, the Internet of the Future will be characterized by a proliferation of smart devices embedded in everyday objects to support daily life activities and a richer user participation in the process of creation and consumption of interactive content (Mayora et al. 2008). This transition will be achieved through the deployment of a wide range of new technological approaches, which include the use of pervasive computing, virtualization, cloud computing, user generated content, service oriented architectures and semantic web.

The natural and continuous Internet evolution raises some questions regarding its appropriate architecture, how to design its building blocks, and the impact and challenges of these changes. This topic will outline some relevant collections of technologies and concepts that are on the path to achieve the Future Internet, such as Web 3.0, Clouding Computing, Internet of Things, and 3D Internet. Moreover, we intend to show how these collections of technologies and concepts can help building some of the key features described in the previous topic that will serve as basis for the Future Internet.

3.1 Web 3.0

Web 2.0 represented a social revolution and a shift from professionally published web content to user generated web content. Web 2.0 applications are largely based on mashups that occur at the data level, rather than application level, and often involve the read-write nature (Hendler 2009). The growth of websites that encourage users to interact and produce content such as wikis, blogs and social networks is an example of this phenomenon. As a result, a large amount of unstructured multimedia data is generated across a multitude of distributed hosts, requiring new search and navigation mechanisms in order to find and manipulate relevant information.

In this context, Web 3.0, also called Semantic Web, was conceived to provide a more structured environment to represent and exploit the information generated by the next Web generation. The Semantic Web combines Web technologies and knowledge representation, which is a subfield of artificial intelligence concerned with constructing and maintaining models of the world that enable reasoning about themselves and their associated information (Lassila and Hendler 2007). Recently, Semantic Web has reached new maturity levels with the standardization of its languages - Resource Description Framework (RDF) and Web Ontology Language (OWL) - and the development of technologies that support them. In addition, Semantic Web technologies are transitioning from universities to large companies such as Yahoo, Oracle and IBM. However, much research remains to be done to ensure that Semantic Web can build flexible and intelligent information systems that could use their representational power to describe things in the real world and "do the right thing" even in unexpected situations (Lassila and Hendler 2007).

The Web 3.0 is closely related to several features described previously on section 2. The use of the representational power of Web Semantic to achieve **metadata identification** is one of the most immediate relations. Besides, this metadata representation scheme can be used in the construction of **context sensitivity and adaptation** mechanisms for the Future Internet. Furthermore, the heterogeneous nature of Future Internet devices requires the creation of Web 3.0 mechanisms to ensure that the **content can be shared among different devices** and accessed in a **ubiquitous user Interface**.

3.2 Cloud Computing

Cloud Computing is an emerging paradigm in which a collection of scalable, shared, virtualized and distributed computational resources is capable of providing infrastructure, software, platform, network, and data storage in the form of services over the Internet with the help of required technologies, such as distributed systems, service oriented architecture, grid computing and virtualization (Pokharel and Park 2009). Cloud architecture generally consists of three main divisions: IaaS, PaaS and SaaS (Mell and Grance 2009). Infrastructure as a Service (IaaS) delivers hardware capabilities, such as storage, network and processing (e.g., Amazon Elastic Compute Cloud and Amazon Simple Storage Service). Platform as a Service (PaaS) offers an integrated environment for applications deploy (e.g., Google's App Engine). Software as a Service (SaaS) provides a complete application as a service. Customers of Cloud Computing can rent resources in the form of services over the Internet from third-party providers in order to deploy their applications, and pay based on their demand, instead of incurring huge capital expenditures to build their own hardware or software infrastructures, whereas providers can leverage existing investment to make profits by increasing utilization via multiplexing the resources among the users (Sheu et al. 2009).

Cloud Computing technology's emergence raises questions and points out significant challenges. One of main challenges is the lack of established interoperability standards; the cloud community must build standards to support service discovery for different types of clouds (nowadays customers have to perform manual search) and the use of the same tools (e.g., management tools and virtual server images) with a variety of providers and platforms in order to enable cloud infrastructures to evolve into a worldwide, transparent, portable and flexible platform (Dikaiakos et al. 2009). Another significant challenge is the development of mechanisms that ensure data confidentiality, integrity, and availability; cloud service providers must offer capabilities that include a trustful encryption schema (to ensure that the shared storage environment safeguards all data), an astringent access control policy (to prevent unauthorized access to the data) and safe data backup and storage (Kaufman 2009). Additionally, it is important to investigate new data-protection mechanisms to ensure data privacy and resource security as concerns regarding user privacy and protection against security threats might happen in resource provisioning and during distributed application execution (Dikaiakos et al. 2009).

Cloud Computing technology is a powerful enabler of some of the key features described previously. For instance, it can support **mobility** in Future Internet by supplying computing power and enabling users to access Internet services over very lightweight portable devices rather than through a desktop PC (Dikaiakos et al. 2009). Besides, it is extremely oriented to the **service paradigm**, being able to provide infrastructure, platform and applications as services. Moreover, it will be much easier for Future Internet **users to generate their own services and content** making use of Cloud Computing resources.

3.3 IoT: Internet of Things

The increased computational power of mobile devices, the ubiquity of wireless networks, and the development of modern wireless sensor technologies leads to the exciting vision of the interconnected smart everyday objects, forming a new "Internet of Things" (Baker et al. 2009). The Internet of Things (IoT) can be defined as a worldwide communication network of uniquely addressable and interconnected real-world smart objects based on standard communication protocols (Stuckmann and Zimmermann 2009). The Internet of the future comprises this Internet of "intelligent objects" and "smart environments" and, also, the existing and evolving Internet of Services and Information.

In the near future, it is expected that smart objects used to capture and interpret events from users and environments will be characterised by a high degree of autonomous data capture, event transfer, network connectivity and interoperability (Botterman 2009). However, new approaches have to be found to ensure these objects can provide the necessary intelligence to perceive, understand and actuate in their environments. As users will have their lives continually mapped in computer systems at a very fine level of detail, one of the main challenges of the IoT is how to address regulatory and ethical issues regarding trust and privacy (Van Oranje et al. 2008). Thereby, it is imperative that the mass of data collected by smart objects is used in ways that protect users' privacy and allow them to be in control of their data, ensuring the creation of a trust environment (Hourcade et al. 2009). Moreover, with the increasing usage of bandwidth-intensive applications and the proliferation of wireless-enabled objects, a key concern is how to find efficient ways to use valuable resources, such as broadband bandwidth, radio spectrum and energy.

The Internet of Things is an essential part of the Future Internet and many features described in topic 2 are also related to IoT. For example, making sensor-generated information usable as a source of knowledge will require semantic integration and **metadata identification** that allows different applications to talk with each other and use the same representation for real world knowledge (Botterman 2009). Besides, the IoT will be characterized by a much higher level of heterogeneity than the current Internet as it includes different objects (e.g., distinct functionalities, technologies, and application fields) in the same communication environment (Stuckmann and Zimmermann 2009). This will require the creation of mechanisms that enable the users to **share content among different devices** and can support **scalability** issues, such as addressing and routing.

3.4 3D Internet

The Internet is evolving towards providing richer, personalized and immersive user experience due to the advances in 3D processing software along with the innovations in 3D graphics and computational equipment. These advances give rise to innovative applications and services in areas such as gaming, virtual worlds and communications (European Commission 2008). Although the concept of 3D Internet seems incremental in the sense that it adds 3D graphics to the current Web, it is revolutionary because it provides a complete virtual environment that can facilitate communication, business, and entertainment on a global scale (Alpcan et al. 2007). In fact, the Future 3D Internet should fit some characteristics related to its contents (e.g., 3D, haptic, interactive, intelligent, real time, cross modal, publicly opened, and collaboratively edited/filtered) and its network (e.g., based on content/service-centric model, able to transport 3D multimodal media, scalable and self-adaptable to heterogeneous devices, real time, context-aware) in order to support these new applications and services without disturbing the normal content delivery (Daras and Alvarez 2009).

One of the main challenges for Future 3D Internet is the development of powerful terminals for 3D visualization that provide immersive media for the user everywhere and at any time which requires research for more efficient coding, streaming, broadcasting and visualization in different kind of terminals (Calic et al. 2008). Thus, 3D Internet will carry a significantly greater volume of data and increase the reliance on graphics and interactivity, so it is also crucial to minimize the latency that clients observe when interacting with servers (Alpcan et al. 2007). In addition, it is essential to provide alternatives for enabling the seamless user controlled identity management, ownership and trading of virtual digital objects, and right of use (Zahariadis et al. 2008).

The 3D Media Internet will play a significant role in achieving the key features of Future Internet. For example, the augmented virtual worlds, the collaborative platforms and the moving holograms created in 3D Internet will originate new requirements in terms of information representation and **metadata identification** (Zahariadis et al. 2008). Moreover, the new services and applications will place new demands on **ubiquitous user interfaces** that will have to support novel input (e.g., 3D position sensors), display (e.g., 3D displays) and presentation (e.g., augmented reality) modalities in different kind of terminals (European Commission 2008).

3.5 Considerations

The technological concepts presented are by no means exclusive; there are other ones that can be derived from them, such as virtualization. However, all these technologies need to follow a standard in order to satisfy the desired interoperability levels. Open standards are the most acceptable solution, but they need time to be developed due to its process of approval and homologation. Thus, there are considerable possibilities that proprietary standards will be used while open standards are under development and/or approval. There are researches to define standards and best practices in Cloud Computing, IoT and Web 3.0 (CSA 2009) (TheOpenGroup 2008).

4 Technological Issues for Internet Evolution

In order to achieve a Future Internet based on the concepts previously seen, it is necessary to go beyond current technologies, not only creating new technologies but also integrating them with current ones in a user-centric way. We have identified 4 key technological issues, each of them with several implications that need to be addressed in the evolution from the Current Internet to the Future Internet.

The first issue is **Device Connectivity**. Currently most Internet-connected devices are servers and desktops using wired connections, although wireless connections have been growing as the number of wireless-enabled devices (e.g. IEEE 802.11 notebooks and netbooks, 3G smartphones) increase. In an IoT scenario however, the number of wireless connections easily outgrow wired ones, leading wireless connectivity to a predominant position in the Future Internet and leaving wired connections mainly for infrastructure for backbones and provider-to-home.

Devices may use multiple connections to the Internet (e.g., it is not uncommon nowadays to have a smartphone with WLAN and cellular network connectivity). Also, IoT requires several types of connection in order to connect "things" in a useful way. This need for multiple connections and the large number of devices that need to be connected to the Future Internet demand a new, larger addressing space than the one provided by IPv4. One possibility is the use of IPv6; however other novel approaches may be required (eg.: IoT locality awareness may have implications on the way devices are connected and addressed).

Another important issue is **Client Mobility.** At present Mobility is mainly provided by Mobile Network Operators (MNOs) through their GSM or 3G networks, which is still expensive and bandwidth-limited. Due to that devices are mostly static (desktops) or nomadic (notebooks migrating among wireless hotspots). While lowering costs and increasing bandwidth may fulfill current Internet mobility needs, there are other needs that need to be addressed for the Future Internet. One example is device identification: under these kind of networks identification is limited due to how the technology works (e.g.., it is not possible to access a mobile phone directly using an IP address as we can do with computers), so it is necessary to deal with MNOs` services that are not usually available. Besides, mobility options in IPv4 are limited, and IPv6 is still not popular on wireless devices or wireless sensors. Furthermore, there are different mobility types for different scenarios, e.g., Network-based (PMIP, 3GPP-LTE), Host-based (MIP/NEMO, WiMAX), Local (MANET). Thus, mobility in Future Internet should allow bidirectional communication among devices, sensors, and computers. Mobile clients should access IoT and Cloud Computing services without interruptions due to mobility issues.

The third key issue is regarding **Application Development:** current Internet applications are mainly developed by organizations or open source communities. Despite all developers' effort to create customizable applications, it is still hard for a standard user to create its own applications according to his or her wishes. In the Future Internet the task of creating an application is even harder, as new applications need to compile the information acquired by several devices and information stored in regular networks to provide the expected results. Additionally, they require the use of different criteria for processing information such as locality awareness, user preferences, and legal regulations. Therefore, without new development approaches, it will be overwhelming for a standard user to create Future Internet applications.

Applications as we know today will still exist in Future Internet and should be used as resources for other applications/services. However, companies and open source communities need to spend a considerable effort to create "building block" applications and Mashups Services. "Building block" applications need to allow a user to collect and process information of his/her devices or devices around him/her using locality awareness independently of Internet connections. On the other hand, Mashup Services should allow the combination of a large range of services available in the Internet in order to create new services or to enable the interaction with the "building blocks" applications. The Amazon Simple Queue Services (Amazon 2010) is an example of how services can be arranged like Mashup Services. There is also an effort to create "building blocks" as the "Software Fabric" developed to build IoT applications (Rellermeyer et al. 2008).

The final key issue is **Security Provisioning.** Digital certificates and PKI (Public Key Infrastructure) are the base for technologies deployed for security provisioning in the Current Internet. It essentially relies on a hierarchical structure to provide a third part trust entity in which users and providers can trust. However, the Future Internet requires an easy and dynamic way to identify security levels allowing the connection to a wide range of devices, sensors, services, and applications. In addition, it is necessary to meet some requirements of high level privacy while enabling user/devices to easily share information/resources with other users/devices. Thus, there is a necessity of hierarchical and distributed security solutions that allow users to trust in other users, taking into account the behavior, the identification, the locality awareness, and other relevant information.

The addressed issues help understand the technological challenges to be faced on the migration to the Future Internet. In short, the technological challenges on the network infrastructure lies on providing mobility to a wide range of users (computers, devices, sensors, peripherals, etc.) and network connection performance with adequate QoS levels. Besides, the user-driven approach will challenge the applications developers to change the commercial client-server paradigm to a commercial hybrid between client-server and peer-to-peer. Moreover the developers of the "building-blocks" and Service Mashups should (transparently to the user) interact with the network infrastructure in order to verify and allocate network resources according to the components used by the user to create its application/service. Finally, the amount of shared information (generated by both users and network) that needs to be processed by the large number of heterogeneous devices will demand a complete new set of security solutions to provide privacy and adequate security levels.

5 Non-technological Challenges

Although there are several technological challenges to achieve the key features on the Future Internet, there are also challenges related to non-technological issues

(Hausheer et al. 2009). These challenges are related to management and usage of the Future Internet, such as government regulations, business models, and social behavior. Thus, challenges in Future Internet can be classified in four domains (Leva et al. 2009): Political, Economical, Social and Technological. While Technological challenges have been explored previously, the other 3 domains need further clarification.

On the Political domain dependency on the Internet tends to increase continuously, changing the way people, organizations and governments interact. The expected growth of mobile devices requires new government approaches for the regulation of radio spectrum usage and correlated infrastructure. There are also regulation necessities related to the criteria for energy consumption and sustainability.

Regarding the Economic domain, the creation of users' applications using "building blocks" and Service Mashups require new business models. In the last decades it has been adopted a business model paradigm based on profits from server usage and software/content licenses. This model is adequate for specific applications using a pre-defined content and service types, as it provides easy billing and accountability. However, the new paradigm of user-created applications will encourage the development of globalized applications by the use of components from different parts of the world to build it. Applying traditional accountability and billing services on this type of application demand several transactions for the different components, creating a complex scenario for the user. Thus, business models should allow building applications easily while dealing in the background with issues about accountability, pricing, borders, and government regulations. In particular, the Future Internet needs a business model capable of dynamically identifying the resources used by the user to create his/her application/service, making the billing and accountability according to the locality awareness (regulatory laws) of user and used resources, and allowing the user to publish&sell his/her derivate applications/services.

On the Social domain, applications and content generated based on user-driven approaches will stimulate social networks due to the ease of data computing and delivery to the interested person/group. The amount of information generated from a person by his devices/sensors will instigate new social networks and will produce stronger relations (e.g., an application that can show on a map the friends and/or the business contacts in one's vicinity).

Future Internet will not only connect people, but it will also allow better understanding of their needs and behavior. These needs could be indicated by their interest in particular sets of information. It also provides a different perspective about the world, since people are creating their own applications, services, and data. Thus, people can take better decisions using data created by themselves for their own problems. However, there are still problems of resource misuse and information misinterpretation. The human values must be reinforced in order to avoid the misusage of these powerful resources. Furthermore, technological mechanisms will be necessary to allow the identification of these misuses and potential threats.

In addition, some scenarios may impose challenges that are a combination of the previously addressed challenges. For instance, sustainability issues are composed by political, economical, social, and technological challenges. Nowadays users buy a large number of devices (e.g., computers and mobile phones) that are not frequently used and it may result in a considerable amount of Electronic Waste (E-Waste). Thus, Future Internet should allow an easy way to share these rarely used devices to avoid users to buy devices that are not necessary to use several times. The total amount of power consumption can also be reduced if users share their devices, avoiding several devices to obtain the same data, e.g., it is not necessary to have the same meteorological device in each house in a same square. Furthermore, Future Internet can be used to track hardware and verify if a hardware identified as E-Waste in some place can be useful somewhere else.

From these non-technical challenges in Future Internet, one of the biggest issues is how to deal with regulatory laws of several countries. There are several commercial restrictions that must be taking into account and depend on government decisions, e.g., Apple App Store cannot sell games, music and TV shows in some countries due to lack of regulatory laws. Some countries or economic blocks have already defined ways to solve questions related to borders and governments regulations. The European Union has the Trade Barriers Regulation (TBR) which is a trade policy tool that allows EU companies to formally request the European Commission to start an investigation into trade barriers in third countries. Here the challenge is to create an international agreement that allows using the user-driven approach without major restrictions. Thus, the government regulation of the countries may determine how powerful the Future Internet will be.

6 Considerations

Internet has become a very important tool for our daily activities and its use has spread through different social classes. It has impacted on the way we live, on our relationship with other people, on how we work, study and even spend our leisure time. Its evolution has been pushed by different technologies as mobile and fiber optics networks, sensor networks, Semantic Web, visualization techniques, computing system architectures, human-machine interfaces, among others. Today, we are always connected using electronic devices of different sizes and types, different types of networks and different specific applications, trusting on the reliability, security and privacy of this infrastructure and services and being everyday more eager for new features and services.

Due to the significant importance of this scenario several researches around the world are being developed, including FIRE (FIA 2010) in Europe, GENI (GENI 2010) and FIND (FIND 2010) in USA and Future RNP in Brazil. These projects propose new solutions for the challenges originated by the continuous evolution of Internet.

However, as Internet and technologies become even more embedded in our daily tasks, we have less contact with nature and people and, ultimately, we become more distant from ourselves. As Plato has already stated in 400 bc, we begin to see our life through images shown a cave, or through images shown in the screen of our electronic devices, very far from the real life (Plato et al. 1992).

References

1. AKARI: Architecture Design Project for New Generation Network, http://akari-project.nict.go.jp/eng/index2.htm (accessed March 2010)
2. Alpcan, T., Bauckhage, C., Kotsovinos, E.: Towards 3D Internet: Why, What, and How? Cyberworlds, 95–99 (October 2007)
3. Amazon. Amazon: Simple Queue Services, http://aws.amazon.com/sqs/ (accessed February 2010)
4. Baker, N., Zafar, M., Moltchanov, B., Knappmeyer, M.: Context-Aware Systems and Implications for Future Internet. In: Future Internet Conference and Technical Workshops, Prague, Czech Republic (2009)
5. Benslimane, D., Dustdar, S., Sheth, A.: Services Mashups: The New Generation of Web Applications. Internet Computing, IEEE 12(5), 13–15 (2008)
6. Botterman, M.: Internet of Things: An early reality of the Future Internet. Workshop Report (2009)
7. Calic, J., Daras, P., Achilleopoulos, N., Panebarco, M., Mayora, O., Stollenmayer, P., Williams, D., Pennick, T., Magnenat-Thalmann, N., Guerrero, C., Pelt, M., McGrath, T., Fuenmayor, E., Papapulakis, N., Alvarez, F., Kalapanidas, E., Shani, A., Le Moine, J.: User Centric Media of the Future Internet. Next Generation Mobile Applications, Services and Technologies. NGMAST, 433–438 (2008)
8. CSA. Security Guidance for Critical Areas Focus in Cloud Computing, http://www.cloudsecurityalliance.org/guidance/csaguide.pdf (accessed January 2009)
9. Daras, P., Alvarez, F.: A Future Perspective on the 3D Media Internet. In: Towards the Future Internet: a european research perspective, pp. 303–312. IOS Press, Amsterdam (2009)
10. Dikaiakos, M., Katsaros, D., Mehra, P., Pallis, G., Vakali: A Cloud computing: Distributed internet computing for it and scientific research. Internet Computing, IEEE 13(5) (2009)
11. ETSI, Shaping the World: a scientific view. Wiley, UK (2009)
12. European Commission, Research on Future Media and 3D Internet. White paper. Networked Media Unit, Information Society and Media (2008)
13. FIA: Future Internet Assembly, http://www.future-internet.eu/ (accessed March 2010)
14. FIND: Future Internet Design, http://www.nets-find.net/ (accessed March 2010)
15. GENI: Global Environment for Network Innovations, http://www.geni.net/ (accessed March 2010)
16. Gluhak, A., Bauer, M., Montagut, F., Stirbu, V., Johansson, M., Vercher, J., Presser, M.: Towards an Architecture for a Real World Internet. In: Towards the Future Internet: a european research perspective, pp. 313–324. IOS Press, Amsterdam (2009)
17. Hausheer, D., et al.: Future Internet Socio-Economics - Challenges and Perspectives. In: Towards the Future Internet - A European Research Perspective, pp. 1–11. IOS Press, Amsterdam (2009)
18. Hendler, J.: Web 3.0 Emerging. Computer 42(1), 111–113 (2009)

19. Hourcade, J., Neuvo, Y., Wahlster, W., Saracco, R., Posch, R.: Future Internet 2020: call for action by high-level visionary panel. In: European Commission, Information Society and Media, Belgium (2009)
20. Lassila, O., Hendler, J.: Embracing "Web 3.0". Internet Computing, IEEE 11(3), 90–93 (2007)
21. Leva, T., Hammainen, H., Kilkki, K.: Evolving Internet. In: First International Conference on Evolving Internet, INTERNET 2009, pp. 52–59 (2009)
22. Kaufman, L.: Data Security in the World of Cloud Computing. IEEE Security and Privacy, 61–64 (2009)
23. Marcus, A.: Universal, ubiquitous, user-interface design for the disabled and elderly. ACM, New York (2003)
24. Mayora, O., Daras, P., Panebarco, M., Achilleopoulos, N., Stollenmayer, P., Williams, D., Magnenat-Thalmann, N., Guerrero, C., Pelt, M., McGrath, T., Fuenmayor, E., Salama, D., Alvarez, F., Kalapanidas, E., Shani, A., Le Moine, J.: User centric media in the future internet: trends and challenges. In: Proc. of the 3rd International Conference on Digital Interactive Media in Entertainment and Arts. DIMEA 2008., vol. 349, pp. 441–446. ACM, New York (2008)
25. Mell, P., Grance, T.: Draft NIST Working Definition of Cloud Computing. NIST (2009), http://csrc.nist.gov/groups/SNS/cloud-computing/cloud-def-v15.doc (accessed August 2009)
26. NSF. GENI: the Global Environment for Network Innovations. National Science Foundation (2010), http://www.geni.net/ (accessed February 2010)
27. Plato, G., Reeve, C.: Plato: Republic, 2nd edn. Hackett Publishing Company (1992)
28. Pokharel, M., Park, J.: Cloud computing: future solution for e-governance. In: Proceedings of the 3rd International Conference on Theory and Practice of Electronic Governance ICEGOV 2009, November 10-13, vol. 322. ACM, Colombia (2009)
29. Rellermeyer, J., et al.: The Internet of ThingsThe Internet of Things. Cap. The Software Fabric for the Internet of Things, 87–104 (2008)
30. Schonwalder, J., Fouquet, M., Rodosek, G., Hochstatter, I.: Future internet = content + services + management. IEEE Communications Magazine 47(7), 27–33 (2009)
31. Schweitzer, C., Carvalho, T., Ruggiero, W.: A Distributed Mechanism for Trust Propagation and Consolidation in Ad Hoc Networks. In: The International Conference on Information Networking 2006 (ICOIN 2006), Sendai/Japan (2006)
32. Schubert, L., Kipp, A., Wesner, S.: Above the Clouds: From Grids to Service-oriented Operating Systems. In: Towards the Future Internet: a european research perspective, pp. 238–249. IOS Press, Amsterdam (2009)
33. Sheu, P., Wang, S., Wang, Q., Hao, K., Paul, R.: Semantic Computing, Cloud Computing, and Semantic Search Engine. In: Proceedings of the 2009 IEEE International Conference on Semantic Computing (ICSC), pp. 654–657. IEEE Computer Society, Washington (2009)
34. Silva, M., Carvalho, T., Silveira, R., Ferreira, G., Ruggiero, W., Fragnito, H., Waldman, H., Ruggiero, C., Lopez, L.: Fiber-Based Testbed Architecture Enabling Advanced Experimental Research. In: The 2nd International IEEE/Create-Net Conference on Testbeds and Research Infrastructures for the Development of Networks and Communities (TridentCom 2006), Barcelona, Spain (2006)
35. Stuckman, P., Zimmermann, R.: European research on future Internet design. IEEE Wireless Communications 16(5) (2009)
36. The OpenGroup. OSIMM: The Open Group Service Integration Maturity Model - Draf, http://www.opengroup.org/projects/osimm/ (accessed December 2009)

37. Tselentis, G., Domingue, J., Galis, A., Gravas, A., Hausheer, D., Krco, S., Lotz, V., Zahatiadis, T.: Towards the Future Internet: a european research perspective. IOS Press, UK (2009)
38. Van Oranje, C., Krapels, J., Botterman, M., Cave, J.: The future of the internet economy; a discussion paper on critical issues. Technical Report. RAND Europe (2008)
39. Zahariadis, T., Daras, P., Laso, I.: Towards Future 3D Media Internet. In: NEM Summit 2008, Saint-Malo, France (2008)

Challenges for the Brazilian Green and Yellow Internet

Djamel F.H. Sadok, Judith Kelner, and Joseane Fidalgo

Center for Computer Science, The Federal University of Pernambuco. 1235,
Avenue Prof. Moraes Rego, CEP – 50670-901, Recife, Brazil
{jamel,jk,jbf}@gprt.ufpe.br

Abstract. This chapter describes the challenges for designing a new generation network architecture for Brazil. One that not only considers the introduction of the new technological breakthroughs, though important, but also develops scenarios for their deployment, benefiting the remote and poorer regions of Brazil, often forgotten from major development initiatives. The next Yellow and Green Internet (YGI) represents a special opportunity to ensure that our communities do not miss a unique chance to adapt the next Internet to some of the harsh realities of a large part of the Brazilian realities including regions and communities where communications remain almost non-existent and isolation has reigned for considerable time.

1 Introduction

Today's Internet was not designed to become the core for a world-economy, a global information system, a social place, among other emerging usage scenarios. Nevertheless, it undeniably represents a critical infrastructure for societies across the world, connecting people, groups, enterprises, schools, industries, and devices. It remains the scene for continuous great efforts to patch it up sufficiently attempting to attend and adapt to changing society requirements in the form of new applications, services, and technologies. The role of the Internet in the social and economic development is beyond questioning as it has influenced the lives of everyone. As such, the future Internet must not be seen as a mere technical effort, but as a broad global enabler of a future networked digital society.

When reviewing recent technological developments, one cannot help but to think that the next Internet is closer than one may expect as the technology to support is readily available. Advances have been made in many subjects, to name but a few: cross layer design, naming schemes, new structuring (e.g. using turfs, federations, contexts and societies), new operation modes (disruptive, delay tolerant, P2P, with willingness to learn, willingness to share information), dynamic network and service creation and composition, network coding to optimize transport, support for distributed hash tables for fast information retrieval, dynamic negotiations, virtualization, service brokers and orchestration, content-based routing, location information, privacy, autonomic management, distributed simulation, remote instrument control, support for different business models, data centers and could computing, etc.

Nonetheless, there is the danger that yet a clear vision has not been amply discussed. Although we do not need to agree on a single view, we need these to be clear ones. It is important to establish specific and unambiguous requirements and subsequently methods that measure the benefits across all its user communities, including doctors, administrators, lawyers, engineers, students, teachers, governments, businesses, favela[1] dwellers, physically challenged users, farmers, public servants, etc. For each user class, one must ask the question: how would it benefit from a next Internet? The aim is not to invent a bigger Internet but a more flexible one[2]. The YGI[3] effort looks at minimizing the negatives in a large socially divided country such as Brazil.

1.1 Local Drivers

There are already a number of isolated efforts by the local and federal governments to take the Internet to areas with unprivileged school children and the rural and isolated population. For example, the figure bellow shows a mobile or itinerant school bus (*escola itenerante*) offered by the City of Recife in the north East of Brazil, a region that is relatively poorer than others in the country. These are currently around 6 busses that regularly travel to under-privileged areas of the metropolitan region to offer school children the opportunity to take part into the Internet revolution []. Typically, each of these buses carries 13 connected computers, a printer, a scanner, a blackboard, a TV, a video and a radio, see **Erro! A origem da referência não foi encontrada.**.

Fig. 1 Mobile School Bus in Brazil [1]

---

[1] Refers to large and poorer popular residential areas usually surrounding Brazilian metropolitan cities.
[2] Henry Ford was reported to say that had he consulted the general public, he would have made a bigger horse.
[3] The first Brazilian flag was similar to the US flag except for using yellow and green instead.

Similarly, in the Amazon, a region known for its isolation due to the lack of roads and where rivers are the main transport medium, the federal government in cooperation with state and municipal government offers a number of school boats. Further, organizations such as the federation of the industries currently give technical training using boats to residents of the Amazon region as shown in [1].

The rivers of the North are the only transport medium available to the local population. The ministry of education has therefore initiated the construction and deployment of between 200 and 300 catamaran boats, each with a capacity for 20 to 35 students. These have been available by the start of the school year 2009 [2]. These boats are used when the student trip to school is relatively long enough for them to offer additional enforcement classes with access to advanced communications and the Internet [3]. A separate program called "balsa escola" offers boats equipped with four laboratories for teaching the courses on: tourism and hospitality, health services, personal care (hair cut and related services), computer science courses. These boats are equipped with 21 personal computers and operate between 08:00 and 22:00. Experience has shown that the demand for their services is constantly increasing. In addition, they offer other courses, talks, and make presentations according to the needs of the local communities and has already been to the villages: Parintins, Maués, Barreirinha, Nhamundá and has left towards Borba in the second week of May 2009 where it is expected to spend 6 months there.

In a separate initiative, the Brazilian government is looking to spend as much as R$100 millions on building 3.3 thousands school boats to help children in the remote areas of the North. This problem is known as "Caminho da Escola" or the Way to School.

2 The YGI Approach

Researchers in emerging economies are understandably influenced by tendencies and topics selected in the first world and often tend to take their established approaches and results as granted. YGI attempts to change this, asking our community to think, and think locally first.

Brazil enjoys the 5th biggest territorial extension. Such huge geographical area makes Internet availability restricted to highly populated regions. There is a need to reach out for remote and isolated population pockets through different means capable of conveying information. Consequently, YGI adopts design principles that take into consideration:

- The integration of Postal Services, terrestrial transport systems, and existing Disruptive or Delay Tolerant Networks (DTNs);
- Government School Buses and Boats, private river boats crisscrossing long rivers, public transportation systems such as local and regional buses and trains;
- The large vehicular and transport fleet for the implantation of vehicular networks (VANETs) capable of spreading information and knowledge throughout large areas;
- Support the National Security Mission especially in isolated and remote areas.

YGI could be seen as the personal virtual teacher to a school child in a remote area, a rich statistical database by a civil servant, a place where to get specialized help by a farmer, a health consultancy service by a patient, a social gathering place to discuss with others, etc. When content is put onto the network, its author would give YGI the means to semantically classify it. A kind of usage manual would accompany each piece of content as in the example bellow where a statement such as "This content is useful for 2nd grade students in Brazil" shows how the content provider thinks of facilitating the application mining design.

Further, one may think of the deployment of a more powerful communication service in addition to existing ones, whereby, a message may be sent to Joao by stating that the receiver's name is Joao and that he lives 50 km from Macapa in the Amazon region, et voila! things just happen. Such late binding allows the introduction and integration of applications that do not require all the information at once, as is the case today. YGI would allow people who live near a known Internet site, Post Office, a user, a reachable passing boat, etc., to receive and submit information to the Internet in an opportunistic way.

The Brazilian YGI initiative needs, among other goals, to facilitate obtaining statistics on industry, education, user interests, etc. This is currently a costly and lengthy process, prone to mistakes. YGI may inform a user what a hot scientific topic in the area of networking today is. It would create Internet communities to coordinate their activities, deal with local problems, determine policies and decisions, etc. It would also assist farmers intelligently in their activities. For example, it may guide a farmer asking "who can help me export beans from the distant state of Mato Grosso?" We should help reaching regional and rural areas with broadband low cost services using technologies such as cognitive radio, DTN and new routing paradigms. Environment and crisis management have also been a difficult undertaking when dealing with remote isolated areas. Finally, YGI should take E-government nearer to its citizens.

2.1 YGI Scope

In addition to its concern with technological innovation, the Brazilian initiative introduces societal considerations, and shows how YGI may be introduced in mostly remote, economically challenged urban areas and the northern Amazon region of Brazil through two main scenarios:

- The development and evaluation of an integrated delay tolerant opportunistic communication system linking fixed access points at known building, itinerant school boats, land transportation vehicles and vehicular networks (VANETs), satellite services, postal offices, passengers and transport ships, cellular and radio communication systems, and public mobile government administrative and health service itinerant buses.
- The development and deployment of two pilot applications, namely, the virtual educator and the family doctor.

Amazon school children stand to gain a great deal from gaining access to a virtual teacher for a given grade. They should be able to quickly locate and access important and relevant class material especially designed by the virtual teacher with the help of specialized and tailored semantic based smart search engines developed as part of this project. This should also allow isolated schools, teachers and native tribe members to access especially designed virtual libraries, sharing their thoughts with others all over the world including possibly school teachers and communities in Europe, Asia and the Americas.

Similarly, this new network architecture should also enable some basic support for health consultation and the spreading of health awareness, information and education campaign material among the residents of these remote regions. With the wide spread of diseases such as malaria, health care professionals, from Brazil and other countries may engage in relevant research and beneficial cooperation with these communities. New advanced applications that make use of the deployed communication infrastructure will be made available and evaluated on the field. Although there has been a great deal of rapid advances in medicine and technology, rural health care management, mainly in large developing countries such as Brazil, India and China remains unconcerned. The size of investments needed to take advanced health care services to rural areas combined with the lack of telecommunication coverage are seen as limiting factors. YGI looks at the design and evaluation of a low cost solution to improve rural health care management and facilitate disease diagnosis. Telemedicine provides Doctors with the ability to monitor and evaluate treatment effectiveness for patients with visiting difficulty and is particularly useful for healthcare follow-up in rural areas. Information Technology facilitates the distribution of important medical information and knowledge to the medical community and patients. This includes its use by health practitioners to locate useful medical information on the Web. Online services allow physicians access to immediate information and Doctors can use the Internet to exchange complex medical files across the Web. Another major benefit to utilizing e-health care initiatives is that online access allows patients to be better informed about how exactly they can manage their own health, as well as prevent diseases. E-patients are a new breed of patients who are using the Internet to gain specific knowledge about their symptoms and treatments, as well as using the Web to track down nearly every lead they can find on the best type of new treatment. Some of the major IT technologies in healthcare industry include:

- Telemedicine: Uses technologies such as WebTV, smart phones, and wireless devices to interact with patients in their homes. Possible telemedicine services range anywhere from scheduling appointments online, to performing remote surgical procedures directed by a surgeon to a non-surgeon via high bandwidth technologies and video cameras.
- Decision Support Systems (DSS): DSSs can be used to store standard diagnostic techniques for disease management, and can be used as a cross check against a patient's records. For example, clinical decision support systems (CDSS) can be used to send alerts and reminders to patients about preventive care.
  - Information Warehousing and data mining techniques: Pharmaceutical makers are using information warehousing for marketing

purposes whereas Data mining permits health care providers to save costs and provide faster care.
○ Additional applications in the context of rural e-health include medical staff training, news related to new epidemics, and enhanced cooperation among the medics.

Other applications such as home security, itinerant justice system and local and government administrative work, support for elections, etc., are also considered.

3 Technological Drivers

Although seemingly a simple undertaking, a new YGI architecture requires major changes to the current Internet including the evaluation and introduction of the following technologies, many of which are already being discussed in the context of the next Internet are remain unsolved.

3.1 Delay/Disruptive Networking

YGI investigates and proposes a DTN architecture embracing a new class of heterogeneous networks that may suffer disruptions. DTN repercussions affect a large number of architectural design choices including new naming and addressing forms, data encoding and conversion techniques, message and content formats, the way information is routed, security and congestion control, as well as access to existing communication systems such as the postal office system. YGI aims to setup a long running test-bed to ensure that the adopted DTN supports the new services such as the ones described in the previous section.

- Development of new Transport Protocols: for interworking among DTNs and technologies such as sensor networks, satellite and space communications, postal office communications. Although semantically different, such protocol suite should facilitate the interworking of transport protocols such as TCP/IP, raw Ethernet, serial lines, or hand-carried storage drives for delivery. Current approaches suggest the definition of a collection of protocol-specific *convergence layer adapters* (CLAs) that provide the functions necessary to carry DTN protocol data units (called *bundles*) on each of the corresponding protocols [4].
- Develop and Evaluate new Publish and Subscribe algorithms: under this new paradigm there is a need for advanced algorithms capable of matching semantically users preferences to the available published services and content. The magnitude of this problem increases with respect to the number of attributes, preference criteria, support for semantic search, etc [5]. The problem of disseminating information over the next generation YGI must be examined when considering large groups of distributed users in a collaboration system. Scalability is therefore an important design criterion. Further, YGI needs to consider privacy issues relevant to our applications. While a publisher may know

about the set of subscribers, they should be unaware of each other and may only identify the publisher they are receiving data from [6].

3.2 Cognitive Radio

Although wireless communication is a natural candidate in the context of DTN communications, its use presents a number of limitations as shown next:

- The high bandwidth of applications such as e-health and school access requirements may be difficult to meet when using the licensed spectrum as there could be some strong and uncoordinated competition;
- Distances among the different nodes may be large (tens to hundreds of kilometers) hence requiring the use of good propagating spectrum and advanced radio techniques.

Among the major recent breakthroughs, there is a promising technology, that of cognitive radio. It looks at the opportunistic usage of spectrum and is seen as a good candidate for wireless regional communications as shown in the IEEE 802.22 Cognitive TV standard. Currently the spectrum is partitioned statically in command-and-control fashion where the usage semantics and guidelines are defined a-priori. Consequently, the amount of spectrum allocation based on antiquated technologies is often absurdly unnecessary as in the case of using 6 MHZ channels for analog TV broadcast. Further, spectrum allocation is location agnostic leading to the presence of white wholes in many locations. Often big swaths of allocated spectrum are left unused according to recent measurements and several bands (0-3 GHz) are poorly used with varying occupancy levels. In addition, unlicensed spectrum is used indiscriminately leading to interference.

Cognitive Radio is seen as an extension of software defined radio where a wider spectrum is digitally processed. Programmability offers more flexible control over antenna characteristics, band of operation, channel size, power, etc. Adaptation may be made via software download to radio operation including Up/Down conversion, modulation, coding, error correction, MAC etc. In presence of a large Peak/Average variation and large spatio-temporal variation, long term multi-year provisioning of spectrum to meet peak demands does not work. CR has the added capability of sensing its radio environment in order to adequately reconfigure at various layers taking advantage of the time and space diversity. It is important also to mention that off-the-shelf wireless cards, such as the ICS 572 and ICS-554 products [7]. To summarize, Cognitive Radio is seen as the combination of software defined radio, sensing and adaptation. An adaptive radio must be aware of its environment and its own capabilities, operate in a goal-driven autonomous mode, understand or learn how its actions impact its goal and should recall and correlate its past actions, environments, and performance.

CR allows the use of both licensed and unlicensed spectra. PAs a result new business models may emerge where primary (PU - spectrum owner) and secondary (SU-opportunistic) users must share the same radio resources. There are a number of challenges present at both the physical and MAC layers for require further research. For instance, secondary users are required to limit their transmit power to very low levels such that their signal does not interfere with primary users and to implement reliable primary user detection and frequency agility mechanisms. Once a PU is detected, the SU must switch channels and/or apply transmit power control (TPC). There is a need to examine and evaluate sensing algorithms including coherent detection (pn-sequence or pilot), energy detection and cyclostationary feature detection. The selection of radio sensing algorithms must take into consideration factors such as sensing time vs. SNR, their robustness to noise uncertainty and of course implementation complexity.

YGI will develop a number of test-beds including a CR 802.11P prototype to work with collaborative sensing in an attempt to combat fading and local interference, improve sensing reliability and/or reduce sensing time in addition to build a solution that is suitable to deploy on a large scale such as within rural areas. Note that a simple combination of distributed sensing results increases the probability of false alarm, raises new issues for setting detection thresholds and limits the collaborative gain when used in a limited area due to correlation.

3.3 Small Separation Antennas

The next YGI looks at the impact of using antennas with small separation (in the order of lambda/2) to provide un-correlated multipath, however, while minimizing suffered shadow fading to nearby devices. A CR prototype operating similarly to the 802.11 standard with changes to the physical and MAC layers will be built by extending an existing prototype with the use of robust preambles/headers for synchronization and channel estimation, the use of range of modulation/coding options to support different data rates, the use of OFDMA/TDMA, support at the MAC level for dynamic frequency selection, primary protection and coexistence of secondary systems, dynamic Multi-channel operation.

3.4 Opportunistic Communications

DTNs nodes include mobile/fixed wired/wireless devices. These are able to communicate with each other only when they are within transmission range. As a result DTNs suffer from frequent connectivity disruptions, turning their the topology only intermittently and partially connected. There is therefore no guarantee that at any given time one may find an end-to-end path between a pair of nodes. Examples of recent related work include the DieselNet project [8], which features communication devices deployed in a regional bus system, and Pocket Switched Networks (PSNs) [9], which formed by devices that people carry every day, such as cell phones, PDAs, and music players, project Haggle [10]and the European ambient networks project [11].

3.5 YGI New Routing Paradigms

A number of new innovative ideas and protocols have been based on social phenomena such as regional gossiping [12] and rumor routing [13]. Gossip-based models have, for instance, been used in knowledge diffusion work and for computing aggregate values over large networks in order to simplify the task of network control, monitoring and optimization [14]. These are however somehow oversimplified models, as we need to use asynchronous models where there is call for more interaction rather than relying on old acquaintances to lead with the future. In [15], a barter process is assumed between the members that can trade different types of knowledge. Moreover shared knowledge can create a social network trust in order to increase the security of communication between nodes, because a node may not accept information from any unknown one [16],[17],[18], [19].

Biological models have extensively been used to provide reachability, adaptability, scalability and robustness for systems [20], [21], [23]. For example, a Wasp model provides an advanced scheduling scheme for satisfying robustness when confronted with unexpected events as well as having considerably high performance tasks [15], [22]. Ant models have been used for optimized routing, i.e. leading to increase the throughput and reduce the rate of delay in the network [24],[25],[26],[27].

Considerable work has also gone into partially connected networks [16], [28] and epidemic routing (ER) [29]. Such works introduce ER, where random pair-wise exchanges of messages among mobile hosts ensure eventual message delivery [30], [29]. Similarly, Chen and Murphy propose a protocol called Disconnected Transitive Communication which involves the application in locating the node among a cluster of currently connected nodes that it is best to forward the message to [31]. Given that messages are delivered probabilistically in epidemic routing, the application may require the use of acknowledgments. Some optimizations may be further made using techniques such as bloom filters [28] [32].

Relevant ideas may also come from recent work on Opportunistic Routing over disruptive networks (DTN). Here a number of routing strategies have been evaluated including flooding, random walk, replica forwarding with staggered attempts [16], even enhanced link state protocols and hybrid approaches. Zhang in [33] and Small and Haas in [34] studied analytically algorithms derived from epidemic routing. Spyropoulos proposed a multi-copy scheme for DTN routing in [35]. In [36], a DTN routing strategy that minimizes packet loss is developed.

4 Summary of the YGI Brazilian Approach

The next YGI networking technologies, to be validated by this initiative, are strongly required to solve emerging multiple social problems, as well as to have important roles in the future information and networking society. In the table bellow, some expected future network capabilities in each social problem have been listed.

There is a clear case for a separate approach for building the next Internet in Brazil. A clear vision must be set in the minds of the research community to avoid loss of focus on what is relevant to this endeavor. The Akari Japanese new Internet initiative seems inspiring and interesting to cite as one that combines two

Table 1 Future Capabilities by Social Problem

#	Field	Specific Contribution
1	Transport	Fluvial transport network
2	Environment	Prediction, detection, and reporting of environmental pollution
		Detection of environmental changes
3	Disasters	Prediction, detection, and reporting of disasters
		Evacuation and shelter guidance
		Verification of safety and disaster conditions
		Data protection
4	Teaching	Virtual School and Virtual Teacher

equally important goals: 1) maximizing the positives and 2) minimizing the negatives [37]. YGI looks at the second concern that of reducing the information gap the different population segments and helping all of these equally.

New aggressive scenarios, such the ones presented in this text must be raised and discussed. One needs to take a top down approach with the necessary long view. It is at least misleading to assume that such project must be limited to the scientific, business and administrative spheres. One must consult with the layers of the Brazilian society in order to establish a core of clear goals. Next, metrics for evaluating these goals must be established. These should quantify and or qualify the gain reached by the different solutions embedded into YGI. Real testing is needed to measure the impact of each new part of the new Internet. Finally, some global utility function may be developed to model the overall gains for YGI and answer a simple question: did we deliver on our promise?

References

1. http://www.recife.pe.gov.br/pr/seceducacao/escola.html (accessed November 2009)
2. http://portalamazonia.globo.com/pscript/noticias/noticias.php?pag=old&idN=59356 (accessed November 2009)
3. http://www.boatshow.com.br/noticias/viewnews.php?nid=ult8286906d195a82389545466c2c287e51 (accessed November 2009)
4. http://portal.mec.gov.br/index.php?option=com_content&view=article&id=12190&Itemid=86 (accessed November 2009)
5. Fall, K., Farrell, S.: DTN: An Architectural Retrospective. IEEE Journal on Selected Areas in Communications 26(5) (June 2008)
6. Roumani, A., Skillicorn, D.: Mobile Services Discovery and Selection in the Publish/Subscribe Paradigm. In: Proceedings of the 2004 conference of the Centre for Advanced Studies on Collaborative research, pp. 163–173 (2004)
7. Amit, G., Mathur, R., Hall, W., Jahanian, F., Prakash, A., Rasmussen, C.: The Publish/Subscribe Paradigm for Scalable Group Collaboration Systems, Technical Report (1995)

8. http://www.gefanuc.com/products/2052 (accessed November 2009)
9. UMassDieselNet. A Bus-based Disruption Tolerant Network, http://prisms.cs.umass.edu/diesel/ (accessed November 2009)
10. Chaintreau, A., Hui, P., Crowcroft, J., Diot, C., Gass, R., Scott, J.: Impact of human mobility on the design of opportunistic forwarding algorithms. In: Proc. INFOCOM 2006 (2006)
11. http://www.haggleproject.org/index.php/Main_Page (accessed November 2009)
12. http://www.ambient-networks.org (accessed November 2009)
13. Cointet, P., Roth, C.: How Realistic Should Knowledge Diffusion Models Be? Journal of Artificial Societies and Social Simulation 10(3) (2007)
14. Braginsky, D., Estrin, D.: Rumor Routing Algorithm For Sensor Networks. In: Proceedings of the First Workshop on Sensor and Network Applications, pp. 22–31 (2002)
15. Elasity, M., Montresor, A., Babaoglu, O.: Gossip-based aggregation in large dynamic networks. Journal ACM Trans. Comput. Syst. 23(3), 219–252 (2005)
16. Cowan, R., Jonard, N.: Network structure and the diffusion of knowledge. Journal of Economic Dynamics and Control 28(8), 1557–1575 (2004)
17. Vahdat, A., Becker, D.: Epidemic routing for partially-connected ad hoc networks (2000)
18. Bernardos, M., Casar, J., Tarrio, P.: Efficient social routing in sensor fusion networks. In: Information Fusion, 9th International Conference, pp. 1–8 (2006)
19. Marti, S., Ganesan, P., Garcia-Molina, H.: Sprout: P2p routing with social networks. In: Proceedings of the 1st International Workshop on Peer-to-Peer Computing and Databases, pp. 425–435 (2004)
20. Marti, S., Ganesan, P., Garcia-Molina, H.: DHT Routing Using Social Links. In: Voelker, G.M., Shenker, S. (eds.) IPTPS 2004. LNCS, vol. 3279, pp. 100–111. Springer, Heidelberg (2005)
21. Babaoglu, O., Canright, G., Deutsch, A., Di Caro, G., Ducatelle, F., Gambardella, L., Ganguly, N., Jelasity, M., Montemanni, R., Montresor, A., Urnes, T.: Design patterns from biology for distributed computing. ACM Trans. Auton. Adapt. Syst. 1(1), 26–66 (2006)
22. Cicirello, V., Smith, S.: Wasp nests for self-configurable factories. In: Proceedings of the Fifth International Conference on Autonomous Agents, Canada, pp. 473–480 (2001)
23. Cao, Y., Yang, Y., Wang, H.: Integrated Routing Wasp Algorithm and Scheduling Wasp Algorithm for Job Shop Dynamic Scheduling. In: ISECS, pp. 674–678. IEEE Computer Society, Los Alamitos (2008)
24. Cicirello, V., Smith, S.: Insect Societies and Manufacturing. In: The IJCAI 2001 Workshop on Artificial Intelligence and Manufacturing: New AI Paradigms for Manufacturing, pp. 328–329 (2001)
25. Gutjahr, W.: First steps to the runtime complexity analysis of ant colony optimization. Journal Computers & Operations Research 35(9), 2711–2727 (2008)
26. Dorigo, M., Blum, C.: Ant colony optimization theory: A survey. Theoretical Computer Science 344(2-3), 243–278 (2005)
27. Dorigo, M., Maniezzo, V., Colorni, A.: The ant system: optimization by a colony of cooperating agents. Journal = IEEE Transactions on Systems, Man, and Cybernetics Part B: Cybernetics 26(1), 29–41 (1996)
28. Dorigo, M., Gambardella, L.: Ant Colony System: A Cooperative Learning Approach to the Traveling Salesman Problem. Journal IEEE Transactions on Evolutionary Computation 1(1), 53–66 (1997)

29. Lindgren, A., Doria, A., Schelén, O.: Probabilistic Routing in Intermittently Connected Networks". ACM SIGMOBILE Mobile Computing and Communications 7(3), 19–20 (2003)
30. Kenah, E., Robins, J.: Network-based analysis of stochastic SIR epidemic models with random and proportionate mixing. Journal of Theoretical Biology 249(4), 706–722 (2007)
31. Jindal, A., Psounis, K.: Performance Analysis of Epidemic Routing under Contention. In: IEEE Workshop on Delay Tolerant Mobile Networks (DTMN), pp. 539–544 (2006)
32. Chen, X., Murphy, A.: Enabling Disconnected Transitive Communication in Mobile Ad Hoc Networks. In: Workshop on Principles of Mobile Computing, Co-located with PODC 2001, pp. 21–27 (2001)
33. Broder, A., Mitzenmacher, M.: Network applications of Bloom filters: A Survey. Internet Mathematics 1(4), 485–509 (2002)
34. Zhang, X., Neglia, G., Kurose, J., Towsley, D.: Performance Modeling of Epidemic Routing, University of Massachusetts Technical Report CMPSCI 05-44 (2005)
35. Small, T., Haas, Z.: Resource and performance tradeoffs in delay-tolerant wireless networks. In: Proceeding of the ACM SIGCOMM workshop on Delay-tolerant networking, pp. 260–267 (2005)
36. Spyropoulos, T., Psounis, K., Raghavendra, C.: Efficient Routing in Intermittently Connected Mobile Networks: The Multi-copy Case. ACM/IEEE journal of Transactions on Networking (2007)
37. Lipsa, G.: Routing strategy for minimizing the packet loss in disruptive tolerant networks. In: 42nd Annual Conference on Information Sciences and Systems, pp. 1167–1172 (2008)
38. Akari Project Home page,
    http://akari-project.nict.go.jp/eng/index2.htm
    (accessed March 2010)

Author Index

Alberti, Antônio Marcos   79
Alberti, Antonio Marcos   121

de Brito Carvalho, Tereza Cristina Melo   221
Dominicini, Cristina Klippel   221

Fdida, Serge   141
Fidalgo, Joseane   237
Friedman, Timur   141

Kelner, Judith   237

Miers, Charles Christian   221

Ongarelli, Marco A.   57

Parmentelat, Thierry   141
Pasic, Aljosa   205
Pasquali, Nilo   215

Redígolo, Fernando Frota   221
Rego, A.C. Bordeaux   57
Rothenberg, Christian Esteve   57, 121, 179

Sadok, Djamel F.H.   237
Silva, Abraão B.   215
Stanton, Michael   153

Tome, Takashi   57, 121
Tronco, Tania Regina   1, 13, 25, 57, 121

Vogt, Christian   189

Werner, Julius   167